U0084906

Brewers' Handbook

釀製啤酒一點也不困難，不必花大錢購買昂貴的設備，只要幾個簡單的工具和材料，你就能輕鬆在家釀啤酒！

在家釀啤酒

啤酒DIY和啤酒做菜

釀酒專家 錢薇●著

朱雀文化

推薦序/兼具書香、酒香、菜香的釀酒書

錢薇女士曾長期旅居美國加州戴維斯城，那兒的戴維斯加州大學（UC-Davis）設有全美國唯一的釀酒學系。錢女士經歷這個學系的訓練，復經常訪視附近那帕山谷（Nape Valley）及索諾瑪山谷（Sonoma Valley）眾多酒莊，並在自己家裡釀酒自娛及會友。她在四年前回到台灣，適逢我國加入世貿組織、開放民間製酒之際，從那時開始，著實在釀酒技術之傳播及民間酒廠之輔導方面作出不小的貢獻。

國內民間私下釀造米酒或水果酒者很多，啤酒則尚在起步階段，錢女士這本《在家釀啤酒》來得恰是時候。這書對於釀酒學理及各類啤酒的製程有簡單、正確、扼要、清楚的描述，讓初學釀酒者容易上手不會失敗，再附食譜，與讀者分享美食，也更添加了些生活的樂趣。爲這一本兼具書香、酒香、菜香，由具高深專業素養的作者所寫的，有意義、有價值、有巧思的書作序爲跋，本人是很樂意爲之的。

吳瑞碧

2004年4月

於　台大食品科技研究所

A vast resources on brewing, beer tasting and gourmet recipes.

Chien-wei has been living in Davis, California for quite some time now, as UC-Davis has the well-known wine making& brewing classes in the whole country. Ms. Chien has undergone training in enology department and has visited many wineries in the Napa Valley and Sonoma Valley nearby. She started to brew beer at home to entertain herself and her friends. Four years ago, she came back to Taiwan. That was right after Taiwan joined the World Trade Organization and wine making was opened to the public. From that time on, she has contributed to the dissemination of brewing techniques and aided local wine producers.

While there are many locals privately making rice and fruit wine, brewing of beer is just getting underway in Taiwan. This book (Brewers' Handbook) has arrived at the right time. It contains many detailed descriptions of brewing theory and procedures for brewing all kinds of beer. The processes described herein are simple, correct and clear. Beginners should find it easy to handle. And there are recipes at the back to share with the readers and add spice to daily life. Writing a recommendation for this book, which is perfectly written by an experienced, highly-educated expert, offering aromatic brews, is really an honor.

James. S. Wu. phD.

April 2004/10/20

Director Institute of Food Science and Technology

National Taiwan University

自序/在家釀啤酒

「來，喝一杯我自己釀的啤酒！」

只要你有耐心的看完這本書，保證說這句話的人是你！

釀製啤酒一點都不困難，更不必花大錢購買昂貴的設備，家中廚房裡已經有了一部份的設備，再購買幾件必需的工具，加上清潔衛生的概念，仔細看完這本書的時間，就可以輕而易舉的在家釀製啤酒了。

開始釀製啤酒前，先要了解什麼是啤酒：

啤酒和葡萄酒這兩種含酒精的飲料，都是酵母菌在含糖的液體中進行發酵作用，酵母菌把糖吸收消化轉換成能量，進行細胞核的分裂繁殖，同時產生了酒精及二氧化碳二種副產品，糖水就變成酒了。

葡萄酒和啤酒的區別在那裡呢？

葡萄酒是將葡萄中含有的天然糖份，經過酵母菌的發酵作用轉化成酒精（請參考《釀一瓶自己的酒》這本書），而啤酒則是將穀物（主要是大麥）中含有的澱粉，先轉化成麥芽糖，再利用酵母菌的發酵作用轉化成酒精。

釀製啤酒的過程比葡萄酒來得複雜，但是需要的時間比較短。釀製啤酒和葡萄酒時，釀酒師必須要全程掌控所有的步驟，以科學的方法，按部就班的注意每一個細節，才能製作出好喝的成品。

錢薇
聯絡方式
傳真：（02）23583078
e-mail：vector_bio@yahoo.com.tw

Beer Brewing

"Come and have a glass of beer that I brewed!"

If you finish this book with patience and care, I guarantee that you will be the one saying that!

Brewing beer is not difficult at all. There is no need to spend lots of money to purchase expensive equipments. If some of the equipment may already be found in your own kitchen, then just purchase a handful of the really necessary utensils. With correctly cleaning concepts and careful study to this book; you can easily brew great beer at home.

Before we start brewing beer, you have to understand what beer is:

Beer and wine are two different kinds of alcohol drinks. Both are manufactured through the fermentation of yeast in a liquid that contains sugars. Yeast absorbs nutrients from the sugar, transforms it into energy, and then carries on cellular reproduction. The by-products of fermentation, carbon dioxide and alcohol, are produced. Through the magic of yeast, sugar becomes alcohol.

What are the difference between wine and beer?

Wine is made from the natural sugar found in grapes fermented by the yeast into alcohol (see "Make Your Own Wine" as a reference). In Beer, by contrast, the starch in the grain (especially barley) is transformed into malt sugar, and then to alcohol.

The process of brewing beer is more complex than the wine, but required processing time is shorter. When brewing beer or making wine, You have to completely control every procedure using scientific methods and going step by step, paying close attention to every detail, so that a glass of delicious beer will be your rewards.

Chienwei

c o n t e n t s

Chapter I Basic

基礎篇

Chapter II Beer Brewing

釀酒篇

Chapter III

Cooking With Beers

應用篇-用啤酒做中西佳餚

Basic

Chapter I

1

基礎篇

一、啤酒釀製流程圖The process of brewing beer

二、基本設備基本設備Basic Equipment

三、基本材料Basic Ingredients

四、STEP BY STEP輕鬆釀啤酒Brew Beer with Ease

啤酒釀製流程圖

啤酒的釀製過程如下圖所示：

啤酒釀製程序

原料（大麥Barley）

↓

浸水

↓

控溫：發芽室進行發芽

↓

低溫烘焙

↓

研磨成細粒

↓

浸熱水萃取麥芽糖

↓

製成啤酒基本原料麥芽汁（wort）

↓

加啤酒花煮沸

↓

過濾除去殘渣

↓

加入酵母菌，進行發酵

↓

發酵完成，虹吸，澄清

↓

第二次發酵

↓

裝瓶→生啤酒

↓

62°C加熱30分鐘

↓

啤酒

The Process of Beer Brewing

The process of beer brewing is shown in the chart below :

The process of beer brewing

Ingredient (Barley)

↓

Soaking

↓

Temperature controlling: Malting

↓

Baking under low temperature

↓

Crush the grains into fine mash-milling

↓

Soak in hot water ,mashing

↓

Sparging to make the main ingredient of the beer - wort

↓

Adding hops and bring to a boil

↓

Removing and discarding sediment

↓

Pitching in the yeast and starting fermentation

↓

Fermentation finish, siphoning, clarification

↓

The second fermentation

↓

Bottle→Raw beer

↓

62C Heat for 30 minutes

↓

Beer

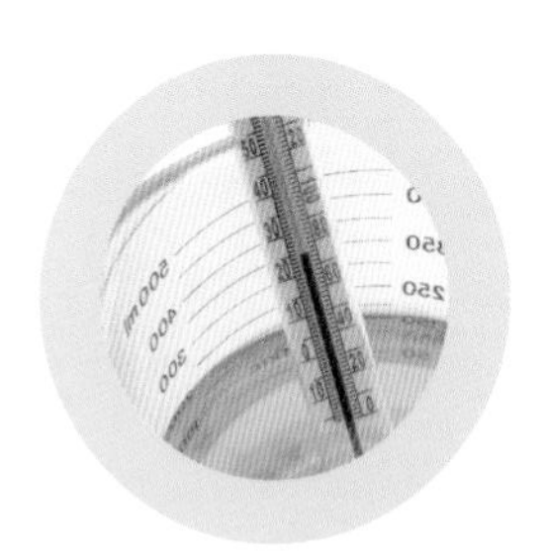

依照啤酒釀製過程，逐項解說如下：

1.原料

啤酒的主要原料是大麥（Barley），大麥發芽時，麥粒會產生神奇的變化，釋放出澱粉轉化酵素，將麥粒含有的澱粉轉化成麥芽糖，同時也會產生蛋白分解酵素把蛋白質分解成氨基酸。麥芽糖和氨基酸是為了滋養即將長出的嫩芽，提供充足的營養讓麥粒長成茁壯的麥穗；釀製啤酒時，麥芽糖變成了酵母菌的營養來源，讓酵母菌充滿能量，快速的分裂繁殖，進行發酵作用，氨基酸豐富了啤酒的香醇及質感。

2.浸水

大麥用水浸泡2～3天，為了讓乾燥的麥粒吸飽水份（Soaking with water）。因為麥粒的表面附有很多的土壤及雜質，浸水時，要記得每天換水，浸泡麥粒的同時，也將麥粒清洗乾淨了，浸泡約3天左右，當麥粒的含水量達到40％左右時，就可以自水中撈出來。

3.發芽

大麥浸飽水份後，要裝入發芽槽中或平鋪在發芽室的地板上。這時，溫度和濕度的控制是大麥發芽是否成功的關鍵。發芽室的溫度要控制在13～18℃之間。慢慢，約3～5天後，麥粒底部冒出了白色的鬚根，同時麥粒釋放出澱粉轉化酵素，將澱粉轉化成麥芽糖，當嫩芽自根部慢慢向上延伸時，要非常注意它的生長狀況。

4. 烘焙（Bake）

嫩芽生長到整顆麥粒2/3的高度時，麥粒所含有的澱粉已完全轉化成麥芽糖，所以要中止麥芽繼續生長。此時將麥芽自發芽室移到烘焙爐中，以80～100℃的低溫慢烘。烘焙有兩個作用：（1）中止麥芽繼續生長。（2）烘焙時不同的溫度、火候和時間，均可焙製出不同風味、口感色澤的麥芽，增加了釀製啤酒的多樣性及變化。依烘焙溫不同，略說明如下：

A.以80℃的溫度烘焙，麥芽呈現淡黃色時即取出，這種程度的烘焙麥芽是最基本的釀製啤酒原料，它可以單獨使用或是滲雜不同的原料混合使用。

B.以90℃的溫度烘焙出金黃色的麥芽，俗稱mild Ale malt，釀出的啤酒帶有琥珀的顏色。

C.高溫烘焙（超過100℃）的麥芽可分為慕尼黑式（Munich）和維也納式（Vienna）兩種。維也納式烘焙麥芽釀出的啤酒，香味濃郁口感豐厚呈現琥珀的色澤。慕尼黑式麥芽釀出的啤酒香味濃郁，顏色呈棕黑色帶有焦糖的芳香。以下3圖為德國麥芽、黑麥芽和比利時麥芽。

德國麥芽（低溫烘焙）

比利時麥芽（中溫烘焙）

黑麥芽（高溫烘焙）

The Process of Brewing Beer

1.Ingredient (Barley)

The main ingredient of beer is barley. When barley sprouts, a miracle happens. Starch-converting enzymes are released, breaking down the starch in the sprouts into malt sugar. At the same time, it produces enzymes that dissolve proteins into their constituent amino acids. The malt sugar and amino acids nourish the young sprouts and offer complete nutrients that enable the barley to grow into stalks of grain. During brewing, the malt sugar becomes the source of nutrients for the yeast, which rapidly reproduces, and carrying out fermentation. Meanwhile the amino acids contribute the aroma and body of the beer.

2. Soaking

Soak barley in water for 2-3 days in order to let the dry barley soak with water. Because there are many soil or impurities attached to the surface of the barley, when soaking, remember to change water every day. So the barley is rinsed during soaking. Soak for about 3 days, when barley contains up to 40% liquid, then it can be removed from water.

3. Temperature controlling: malting in the sprouting room

After the barley has been soaked in water, it is removed to the sprouting chamber or spread evenly over on the floor in the sprouting room. Control of the temperature and moisture is the key to whether the malting will be successful. The temperature of the sprouting room should be between 13 and 18 C. Gradually, after 3 to 5 days, little roots like white beards start appearing at the bottom of the barley. This is a signal that the barley is releasing enzymes that break down the starches into malt sugar. When the sprouts start sending out roots, you must monitor its growth carefully.

4. Bake at low temperature

When the sprout reaches roughly 2/3 the height of the whole Grain, the starch that it contains has been broken down into malt sugar and it is time to stop the sprouts' growth. Now remove the sprout from the sprouting room to the oven, and bake slowly at 80 to 100 C. The baking has two functions: (1) to stop the growth of the sprouts. (2) Through variations in temperature, heat and time, different kinds of flavor and colors of beer will be produced, increasing the diversity and variety of the end product.

Malt color is controlled during the baking by means of the temperature used:

A. Bake at 80C temperature, and remove when the barley sprouts appear light yellow. This degree of baking is the basic brewing ingredient. The sprouts can be used individually or combined with other ingredients in brewing.

B. Barley sprouts baked at 90C are golden. Known as mild Ale malt, the resultant beer also appears to be milky-yellow color.

C.Barley sprouts baked at high temperature (over 100C) are used to produce either Munich style brews or Vienna style brews. Vienna style is tasty, thick, and appears milky-yellow color. Beer brewed in Munich style has thick aroma, and the color is dark brown with caramel aroma. The photo below shows German malt, Belgium malt and black malt .

5.研磨（Milling the malt）

烘焙完成的麥芽，要研磨成細粒，以便萃取出麥芽中含有的麥芽糖，及其它風味物質。但不能將麥芽磨成粉末，因為下一個步驟是要把磨成細粒的麥芽浸泡在熱中裡，若是磨成粉末加入熱水中就過濾不出麥芽汁了。

6.萃取麥芽糖（Mashing）

第3個步驟時，麥粒釋放出澱粉轉化酵素，將澱粉轉化成麥芽糖，而這個步驟就是將麥粒中含有的麥芽糖萃取出來。

（1）將磨碎的麥芽浸泡在溫度50～55℃（122～131℉）的熱水中約1個小時，讓麥粒中釋放出來的蛋白酵素把麥粒中含有的蛋白質轉化成氨基酸，麥芽汁中含有太多蛋白質，會造成啤酒混濁不穩定的狀態。

（2）1小時後浸泡麥芽的水溫加熱到66℃（150℉）左右，繼續浸泡1個小時，這時的作用是利用麥芽中釋放出來的澱粉酵素把澱粉完全的轉化成麥芽糖。每10分鐘要將麥芽用力攪拌，盡量讓麥芽糖溶入熱水裡。

7.過濾出麥芽汁（Sparging）

以66℃熱水浸泡麥芽細粒進行糖化作用1個小時後，先將浸泡麥芽的熱水倒出來，將此麥芽汁加熱到71～76℃（160～168℉），回沖到麥芽細粒，再一次的過濾出麥芽汁，以便將麥芽中的糖份，完全的浸泡沖洗到麥芽汁中。

為什麼設定熱水溫度為76℃呢？因為超過76℃的水太熱了，回沖到麥芽細粒時，不但溶出了糖份，也溶出了麥粒中所含的單寧及尚未轉化成氨基酸的蛋白質，麥芽汁中若含有過量的單寧及蛋白質，釀製出的啤酒會有穀粒的青澀味道，而且呈混濁狀態不易澄清。

8.煮沸麥芽汁（boiling with hops）及啤酒花

當我們進行前面的步驟時，空氣中很多看不見的微生物、雜菌及灰塵也會溶入麥芽汁中，所以要用大火煮沸麥芽汁來殺菌消毒、避免麥芽汁被污染。

麥芽汁只要煮沸就已經達到殺菌的作用，為什麼還要繼續煮1個小時呢？這又是大自然的奇妙作用了，麥芽汁要加入啤酒花才會成為啤酒，具備啤酒特殊的香氣及苦味，啤酒花含有兩種物質，一種是芳香油成份，另外一種是啤酒花脂中特有的α酸及β酸，可為啤酒添加怡人的苦味。

啤酒花中的芳香油成份，只需要幾分鐘就完全溶入熱水中，長時間的煮沸，香味反而會蒸發消失。但啤酒花脂中含有的α酸及β酸，很難溶入水中，需經過長時間的煮沸。啤酒花中含有的脂，本身會產生同性異位化（Isomerize）的改變，它的化學結構不變，但原子排列順序改變，將脂中含有的α酸及β酸釋放出來，溶入沸滾的麥芽汁中；所以長時間的煮沸，讓啤酒花在沸水中上下翻滾是唯一能萃取啤酒花苦味的方法。

就因為啤酒花特殊的性質，釀製啤酒時，要先將適量的啤酒花加入麥芽汁中煮沸1小時萃取啤酒花的苦味，等要關火的前5分鐘，甚至關火時再加入少量的啤酒花，來萃取啤酒花的香味。

5. Milling the Malt

After baking, the grains are milled into small particles to extract the malt sugar avoid grinding it into fine powder. The wort cannot be filtered or sieved if the grains is in powder form.

6. Soaking in hot water to create the wort-sparging

The malting process produced enzymes that convert starch to malt sugar. And this process is extracting malt sugar into hot water.

(1) Soak crushed grain in hot water at 50C -55C (122~131F) for about 1 hour to release enzymes that break down the proteins into amino acids. If the wort contains too much protein, the beer will be under unclear and unstable.

(2) After soaking in hot water for 1 hour, increase the temperature of the water to approximately 66C (150C), and continue soaking for 1 hour longer. The α amylase & β amylase are active in this step. Stir vigorously every 10 minutes to enable the malt sugar to dissolve in hot water as much as possible.

7. Sparging

Soak the barley sprouts in 66C hot water for 1 hour, pour out the hot water and increase the temperature to 71~76C (160C~168F). Pour back into the barley sprouts, then pour the wort out, so that the sugar in the barley sprouts will completely dissolve in the wort.

Why set the temperature to 76C? Because higher than 76C is too hot. When poured back into the barley sprouts, it will dissolve the tannins as well as the proteins. If the wort contains too much tannin and protein, the finished beer will taste bitter and cloudy.

8. Adding hops and Boiling

When we are carrying out the procedures above, there are many micro-organisms in the air, bacteria and dusts will contaminate the sweet wort, therefore, boiling wort over high heat can kill the bacteria, sterilizing to prevent the wort be contaminated.

Boiling the wort also sterilizes it. Hops have to be added to the wort to create the special aroma and bitterness of beer. Hops contain two flavorings, aromatic oils and α-acid and β-acid, which add bitterness to the beer.The aromatic oils in the hops need only few minutes to dissolve in hot wort. A long period of boiling will make the aromatic oils evaporate. However the α-acid and β-acid in hops dissolve in water only with difficulty. They need to be boiled for some time. The oils in hops will isomerize as it boils. The molecular structure remains the same, though the atoms structure changes. It will release the α-acid and β-acid, which then dissolve in the boiling wort. Therefore, a long period of vigorous boiling is the only way to break down the hops, essential oils into the wort.Because of the particular characteristics of the hops, when brewing beer, add a suitable amount of hops to the wort and boil for one hour to release the aroma of the hops, then 5 minutes before removing from heat, or even when the heat is turn off, add a small amount of hops, to release the aroma of the hops.

在使用啤酒花時一定要注意：煮好的麥芽汁透過濾網倒入第一次發酵瓶中，濾出的啤酒花及殘渣需丟棄不要。快速的將麥芽汁冷卻可以減少麥芽汁被氧化的機會，如果麥芽汁中含有過多的溶氧，麥芽汁的顏色會變深，加速啤酒的氧化，造成啤酒混濁不清失去清新的風味，在本章第四篇〈STEP BY STEP輕鬆釀啤酒〉中，我會詳細的說明如何快速降低麥芽汁溫度的方法。

9.添加酵母菌進行發酵（Fermentation）

市售啤酒的種類很多，尤其是歐美各國的超市，各種各樣的名稱真會讓你不知如何挑選，其實啤酒基本上只有兩種，英式啤酒Ale是利用上層發酵的酵母菌釀製而成，德式啤酒Lager是利用下層發酵的酵母菌釀製而成。

酵母菌是單細胞植物，空氣中飄浮著很多野生的酵母菌，而釀製啤酒時一定要使用啤酒專用酵母菌。當酵母菌遇到了糖及空氣，酵母菌立刻恢復活力，分解糖的營養，吸取空氣中的氧，進行神聖的傳宗接代任務，自體分裂繁殖，當糖被酵母菌分解時產生兩種副產品：酒精及二氧化碳，也就是發酵作用。

麥芽汁中的糖被酵母菌吸收分解，含糖量慢慢的減少，但酒精及二氧化碳增加，最後麥芽汁中的糖沒有了，酵母菌沒了補充營養的糖，為了自保，酵母菌停止分裂繁殖，進入冬眠，慢慢沈降到底層，原來漂浮在麥芽汁表面的Ales酵母菌，失去活力後也會沈降到底層，利用虹吸就可以將上層的新釀啤酒和底層的沈澱物分離。

10.澄清（fining）

麥芽汁經過酵母菌的發酵作用，糖被分解產生酒精、二氧化碳，蛋白質被分解產生各種不同的物質與氨基酸，酵母菌更是繁殖快速產生約10^7的數量。當發酵作用完成時，大部份的酵母菌因麥芽汁中的營養（糖）已被消耗殆盡，在缺乏營養的狀態下，為了自保，酵母菌進入冬眠狀態失去活力，紛紛沈降到發酵桶的底部；但麥芽汁中仍含有很多懸浮雜質，為了釀製出清澈的啤酒，利用澄清劑來澄清是很重要的步驟。吉利丁使用方便，是家庭釀製啤酒最常使用的澄清劑。

11.第二次發酵

啤酒的特色是什麼？當你把冰涼的啤酒倒入啤酒杯中，不斷湧出的白色泡沫及入口清涼，充滿了躍動的氣泡，滿足了你的視覺及口感，第二次發酵就可創造出這種效果。

麥芽汁經過酵母菌的發酵作用已經轉化成含有酒精的麥芽汁飲料，當發酵作用完成後，雖然大部份的酵母已經沈降到發酵桶的底部，經過虹吸換桶，但是澄清的新釀啤酒中仍有少許呈冬眠狀態的酵母菌漂浮在新釀啤酒中，此時加入少量的糖，讓酵母菌重新獲得營養補充，恢復活力，酵母菌又開始了它天生的本能，進行發酵作用，產生了酒精及二氧化碳。

When using hops, there is one thing you need to pay attention to: When the wort is poured into the first fermenter through the sieve, the hops and sediments should be removed and discarded.

Cooling the wort down rapidly will decrease the chance of wort being oxidized .If wort contains too much dissolved air, the color will darken as the beer oxidizes, causing the beer to look cloudy and to lose its fresh flavor. In chapter four of this section (Brewing Beer with ease), I will again give detailed methods on how to lower the temperature rapidly.

9. Pitching yeast and starting the fermentation process

There are many kinds of beer available, especially in European supermarkets. It is often not clear what all these different names mean. In fact, basically there are just two kinds of beer, English Ale, brewed with the top-fermenting yeast, and German Lager, brewed with bottom fermenting yeast.

Yeast is a single celled plant, and there are many wild yeast strains floating in the air. Specialized brewer ' s yeast is used in brewing. When yeast comes in contact with sugar and air, it immediately begins carrying on the process of energy production and conversion. It dissolves the sugar's nutrients and absorbs oxygen from the air, then carry out its holy mission of reproduction. When the sugar is dissolved by the yeast, it produces two by-products: alcohol and carbon dioxide. The entire process is called fermentation.

As the sugar in the wort is absorbed by the yeast,, the sugar content gradually decreases, while the alcohol and carbon dioxide content increases. When there is no more sugar in the wort, the yeast stops reproduction and becomes dormant, sinking to the bottom. In ales yeast floating on surface will deactivate and sink to the bottom to become dormant. Now siphon and separate the clear liquid on top from the sediments at the bottom.

10. Clarifying and fining

In this step, the sugar in wort has been converted into alcohol and carbon dioxide through fermentation. The proteins converted into many different minerals and amino acids. Yeast reproduces rapidly. After fermentation the yeast will become dormant and cease reproduction, due to a lack of nutrients. It will settle to the bottom of the fermenting bucket. However, there will still be many floating sediments on the surface of the wort. Thus, in order to brew a clear, light beer, clarifying is a very important procedure. Gelatin , as a fining agent, is used often in home brewing.

11. The secondary fermentation

What are the characteristics of beer? When you pour an ice cold beer into a beer mug, it foams and tastes crisp and cold. The bubbly foam satisfies your sight and taste. The second fermentation creates this effect.

After fermentation, the wort contains alcohol, and the yeast ceases reproduction and sinks to the bottom of the fermenter. Even after siphoning and bottling, there will be a small amount of yeast still floating in the brewed beer in a dormant state. This is the time to add a little sugar to enable the yeast begin reproducing. The yeast will once ferment the sugar to produce alcohol and carbon dioxide.

第二次發酵時，不再像第一次發酵時使用可讓二氧化碳氣體排出的發酵鎖，反而是將第二次發酵中的啤酒裝入壓力桶中，或直接裝入可抗壓力的啤酒瓶中，儲存在低溫的環境下，這時二氧化碳氣體只好溶入酒液中，等到啤酒開瓶時才能恢復自由，逃入空氣，這就是啤酒開瓶時會有那麼多氣泡的原因。

12.裝瓶（Bottling）

生啤酒就是經過第二次發酵處理過的新釀啤酒。

新釀啤酒經過了澄清及第二次發酵處理後，就要裝瓶了。以虹吸管將新釀啤酒裝入啤酒瓶，不要裝的太滿，餘留大約2吋的空間（如圖示）。留的空間不夠，酵母菌沒有足夠的空氣進行發酵作用，不容易產生足夠的二氧化碳氣；而如果餘留了太多的空間，啤酒瓶則可能受不了過多的二氧化碳氣的壓力而爆裂。

新釀啤酒裝入啤酒瓶，需留約2吋空間，讓酵母繼續發酵。
Do not fill it up all the way; instead, leave about 2 inches of space

13.熟成

啤酒裝瓶好立刻加蓋密封，剛裝瓶的啤酒要放在陰暗冷涼的地方熟成一段時間，讓剛加進去的糖，提供給酵母菌恢復活力的營養，再次進行發酵作用，這樣才能產生足夠的二氧化碳。同時，靜置熟成也會增加啤酒的風味及口感，Ale需放置1～2週再喝，Lager則要放置3～5週才會好喝，若是你不喜歡生啤酒。這時你也可以用加熱的方式將生啤酒處理成一般啤酒。生啤酒裝瓶後浸在62℃的熱水中30分鐘，進行穩定處理，就成了一般的啤酒。

看到這裡，你心裡是否會有一個疑問？

新釀啤酒經過第二次發酵處理，讓酵母菌從冬眠狀態再恢復活力進行發酵作用，那麼當第二次發酵作用停止後，產生的沈澱物堆積在啤酒瓶底要如何處理呢？

如果你有這個疑問，恭喜！表示你已經讀通了這本書，有了很正確的釀啤酒概念。

是的，啤酒瓶底經過第二次發酵後，會有一層沈澱物沈積在瓶底，除非用力的搖晃瓶子，否則這層沈澱物並不會讓瓶中的啤酒混濁，啤酒和葡萄酒熟成時放置的方法很不一樣，葡萄酒要側置瓶子放倒，讓酒液浸濕軟木塞以免氧化，啤酒則要正放，瓶口向上直立，目的就是讓沈澱物沈積在瓶底。

喝啤酒時，要準備一個比啤酒瓶容量大一點的啤酒杯，將瓶中的啤酒慢慢倒入啤酒杯中，當沈澱物昇到瓶頸快要倒出來時，立刻停止，這樣你就有一大杯清澈的啤酒了，cheers！

新釀啤酒經過第二次發酵處理，將發酵作用產生的二氧化碳氣溶入啤酒中產生自然的氣泡是傳統的啤酒釀製方法。另外有一個更簡單的啤酒人工充氣法，就是用新釀啤酒加入壓縮二氧化碳氣，就像汽水一樣，立刻新釀啤酒就成了充滿氣泡的啤酒。

At the second fermentation, the fermentation lock that we use in the primary fermentation to release the carbon dioxide is not necessary any more. On the contrary, we bottle the beer from the second fermentation into the pressure bucket, or directly into the bottle. Stored at low temperature, the carbon dioxide cannot release but dissolve into the beer, escaping into the air when the bottle is opened. That is why so much gas foams out when the beer bottle is opened.

12. Racking and Bottling

The sweet wort finished the first fermentation and prepared ready for the secondary fermentation. It is time to be bottling. Use a siphon to bottle the brewed beer. Do not fill it up all the way; instead, leave about 2 inches of space (see photo). If there is not enough space, the yeast will not have enough air to carry out fermentation and produce carbon dioxide. However, if too much space is left, beer bottle might explode due to pressure from carbon dioxide build-up.

13. Aging

After the finished beer is bottled, immediately seal tightly and store in a dark cool places to let it age for a period of time. The added sugar provides the yeast with the nutrients it needs to revive and carry out secondary fermentation, so that enough carbon dioxide can be produced. At the same time, aging will increase the flavor and bitterness of the beer. Ale needs to age about 1 ~ 2 weeks before serving. Lager needs to age about 3~5 weeks to have a tasty flavor. If you do not prefer raw beer, you can heat up the raw beer to transform it into an ordinary beer. After the aging period, place the raw beer in hot water at 62C for 30 minutes to stabilize it. It will then become ordinary beer.

What questions might you have after reading this?

When the beer undergoes secondary fermentation, the yeast revives. After the secondary fermentation ceases, how do you handle the yeast sediments from the fermentation?

If you have this question in mind, congratulations! It means that you already understand this book and have sound intuitions about brewing beer.

Yes, there will be a layer of yeast sediment at the bottom after the secondary fermentation. Unless you shake the bottle strongly, this layer of sediment will not make the beer unclear. After bottling, beer and wine must be stored differently. Wine bottle has to be laid down and so the cork can be soaked in the wine. Beer bottles have to be placed upright to permit the sediments to sink to the bottom.

When serving beer, a beer mug larger than the beer bottle has to be used. Pour the beer gradually in the mug. When the sediments are about to reach the mouth of the bottle, stop right away. This way you have a mug of clear beer. Cheers!

The carbon oxide produced by the secondary fermentation will dissolve into the beer and produce natural foam. This is the traditional brewing methods. There is also a simpler method, which is to pressurize carbon dioxide into the finished beer, just like the soda. The finished beer will then have a full head in no time at all.

基本設備 Basic Equipment

在家釀製啤酒，並不需要花錢買昂貴的設備，先到廚房找找看，再去補足沒有的設備。

Brewing beer at home does not require expensive equipment. Inventory your kitchen first and see what you need to acquire.

煮湯的不銹鋼鍋
（需刷洗清潔沒有油垢）。
1 Stainless steel cooking pot (clean, with no absolutely no oil residues)

標準量匙及量杯及秤
1 Standard measuring spoon, 1 measuring cup and scale

虹吸塑膠管1根（約6呎長）
1 siphon plastic tube (approximately 6 feet long)

長柄湯匙1支
1 long handled stirring paddle

2個25公升的玻璃瓶
（果酒桶），第一次發酵及第二次發酵用。
2 25-liter glass bottle,for first and secondary fermentation.

比重測糖計1支
1 hydrometer

烹飪用溫度計1個
1 cooking thermometer

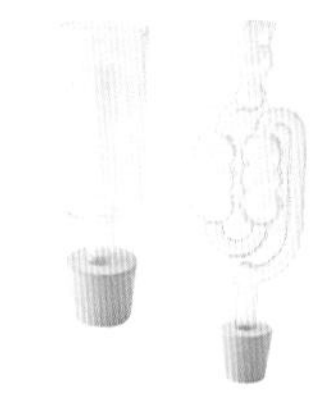

發酵鎖2個
2 fermentation locks

濾網1個
1 sieve

空啤酒瓶2箱
2 cases empty beer bottles

啤酒瓶蓋
Beer caps as needed

消毒殺菌用偏亞硫酸鉀
1 bottle of potassium metabisulfite

洗玻璃瓶刷1支
1 bottle brush for cleaning glass bottle

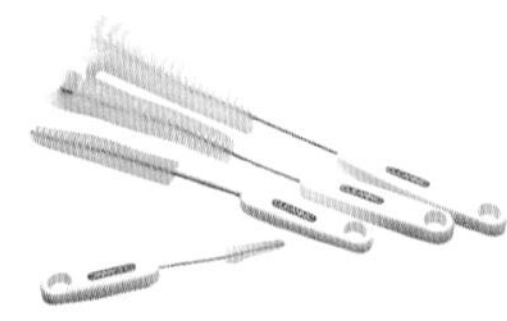

1. 不銹鋼鍋／Stainless steel cooking pot (clean with absolutely no oil residues)

啤酒最怕油垢，有沒有發現，用油膩的玻璃杯倒啤酒，不容易有泡沫產生；所以煮麥芽汁的大鍋一定要涮洗得非常清潔。如果你時常釀製啤酒的話，建議你準備一個煮麥芽汁專用鍋。

It is intolerable to use an oily pot when brewing beer. Have you ever noticed that when you use an oily glass to hold beer, it foams only with difficulty. Hence the pot for cooking the wort has to be extremely clean. If you brew beer often, I suggest you to reserve a pot specifically for cooking the wort.

2. 25公升的玻璃瓶2個／2 25-liter glass bottle, for first and secondary fermentation.

標準的罐裝麥芽汁（約2公斤）一次可以釀製19～20公升的啤酒，所以建議大家準備25公升的玻璃瓶，當第一次發酵時，仍有足夠的空間讓發酵時產生的泡沫不會滿溢出來。

A standard canned wort liquid (about 2 kg) typically yields about 19~20 liters of beer. I suggest that a 25-liter glass bottle be used, so that there is enough space to keep the foam produced during the primary fermentation from overflowing.

3. 長柄湯匙／1 long handled stirring paddle

攪拌麥芽汁及第二次發酵時用來攪拌添加的吉利丁及果糖，使用不銹鋼材質的長柄湯匙，容易清洗，保持乾淨。

It is use for stirring when adding gelatin and sugar during the second fermentation. Use a stainless steel paddle with a long handledas it is easier to clean.

4. 濾網／1 sieve

挑選網眼細密的濾網，再加一層咖啡濾紙，可以過濾出清澈的麥芽汁。

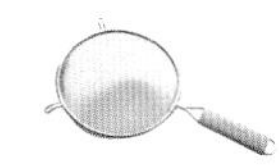

Select one with a very fine netting. By lining with coffee filter paper, a very clear wort can be obtained.

5.標準量匙及量杯及秤／1 Standard measuring spoon, cup and scale

釀製啤酒時份量要計算正確，尤其是啤酒花，需要10～15公克的份量，要有一個能夠秤量出準確份量的秤。

It is important to calculate the right proportions of ingredients, especially hops, which usually is about 10~15 grams. You need a scale that can measure the correct amount.

6.比重測糖計／1 hydrometer

比重測糖計有一個瘦長的塑膠筒用來裝麥芽汁，將有刻度的比重計放入麥芽汁中漂浮，麥芽汁和刻度成弧狀的曲線，查看弧線的最低點刻度的數字就是麥芽汁的含糖量。發酵作用時，酵母菌將糖分解，產生二氧化碳及酒精兩種副產品，糖的含量對酒精的產生有直接的影響，2％的糖（100公克水中含有2公克的糖）大約產生1%（V/V）的酒精。

罐裝麥芽汁稀釋後的含糖量在7～9度之間，也就是說釀製出來的啤酒含有3.5～4.5%（V/V）的酒精。

The hydrometer has a slim long plastic tube for holding the wort. The scaled hydrometer is placed inside in the wort and floats. The readings is taken from a curved line formed by the wort inside the device. The lowest point of the curve shows the sugar gravity of the sugar dissolved in the wort. When fermentation is being conducted, the yeast dissolves sugar to produce carbon dioxide and alcohol. The sugar gravity has a direct bearing on the alcohol content. Each 2% of sugar (100 grams of water contains about 2 grams of sugar) produces about 1% (V/V) of alcohol.

After diluting, canned wort has a sugar gravity of 7~9, and the finished beer will have an alcohol content of 3.5% ~ 4.5% (V/V).

7.發酵鎖／2 fermentation locks/ Airlocks

為了發酵作用特別設計的工具，裝在發酵瓶口上，可以將二氧化碳氣排出，而外面空氣不會進入發酵瓶內，發酵鎖有各種不同型式（見附圖），作用是相同的。

The airlock is designed for the purpose of fermentation. It is attached to the mouth of the fermenter. It release the carbon dioxide out and prevents airborne contamination during fermentation. There are many types of airlocks (see photos), but the function is the same.

8. 虹吸管／1 plastic siphon hose (approximately 6 feet long)

買一截約小指粗細的透明塑膠管，6呎左右就夠長了，當第一次發酵瓶發酵完成後，利用虹吸作用，把上層澄清的新釀啤酒換到第二個玻璃瓶裡，第二次發酵處理完成後亦是利用虹吸作用裝瓶。

The siphon hose is available in several configurations. Purchase one clear plastic hose about the width of your small finger, 6 feet should be enough. After the primary fermentation, use the siphon hose to suck the upper clear beer into the second glass bottle. After the secondary fermentation, the siphon hose is used again to transfer to bottles.

9.烹飪用溫度計／1 cooking thermometer

如果你是用罐裝麥芽汁原料，只有在煮沸的麥芽汁冷卻到24°C左右添加酵母菌時，需要

用溫度計確定麥芽汁的溫度。如果你是用大麥原料自己發芽，烘焙萃取麥芽汁，一支準確的溫度計是必備的工具，尤其是糖化時水溫的控制是很重要的關鍵。

If you use canned wort as an ingredient, pitch in the yeast when the boiling wort cools to around 14C. At this time, a thermometer is needed to measure the temperature of the wort. If you use barley as your main ingredient and brew it from scratch, an accurate thermometer is a necessity. Control of the water temperature during mashing is a crucial in brewing.

10.洗玻璃瓶刷／1 bottle brush for cleaning glass bottle

釀製啤酒的整個過程中，使用的器具及裝備是否完善，以及做好清潔、殺菌的工作是啤酒製作成功的指標。第一次發酵及第二次發酵使用的玻璃瓶務必要刷洗清潔，做好殺菌的預備工作。

Throughout the processes of brewing beer, sterilization and cleanliness should be the watchwords of a successful finished beer. The glass bottles should be cleaned and sterilized during the first and secondary fermentation.

11.空啤酒瓶／2 cases empty beer bottles

新的啤酒瓶使用前仍要用高溫殺菌或是用偏亞硫酸鉀水溶液清洗殺菌，如果是使用舊的啤酒瓶，務必要刷洗清潔，不可有任何的污垢，澈底做好殺菌的工作。

Even new beer bottles have to be sterilized at high temperature or with potassium metabisulfite. If old ones are used, clean thoroughly to prevent contamination.

12.啤酒蓋／Beer caps as needed

啤酒封口時一定要用這種金屬蓋，不可以用葡萄酒使用的軟木塞。

Use metal beer caps for sealing the bottle, do not use corks as with wine.

13. 偏亞硫酸鉀／1 bottle of potassium metabisulfite

高溫熱水是最好的消毒殺菌方法，偏亞硫酸鉀亦有很好的殺菌效果，只是有氣喘的人最好不要接近含有偏亞硫酸鉀的水溶液，以免引發呼吸不順暢。

Boiling water is the best way to kill bacteria, but potassium metabisulfite is also very effective. People with asthma should avoid potassium metabisulfite solution to prevent breathing discomfort.

基本材料 Basic Ingredients

濃縮麥芽汁
Barley malt extract liquid

啤酒花：市售的啤酒花有整朵乾燥的啤酒花、顆粒狀和塊狀啤酒花三種型式，惟第三種塊狀啤酒花很少看得到。
Hops: hops come in three forms, dried cone-like flowers, pieces, and pellets.

濃縮麥芽粉
Dry malt extract (comes in powdered form)

果糖
Fructose

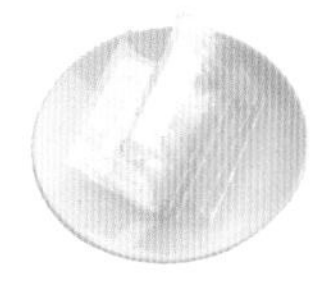

吉利丁
Gelatin

英式啤酒Ale酵母菌
English Ale yeast

德式啤酒Lager酵母菌
German Lager yeast

水
Water

1.濃縮麥芽汁

罐裝濃縮麥芽汁就是將大麥經過發芽作用後，煮出來的麥芽汁（wort）濃縮成像糖漿的濃汁，是初學釀製啤酒者的最愛。

工廠製作濃縮麥芽汁是用傳統大麥發芽的方法，以熱水萃取出麥芽中的麥芽糖，再將甜的麥芽汁低溫真空濃縮，麥芽汁的水份蒸發，成為含水量僅有10％的濃稠麥芽汁，因為低溫真空濃縮的機器設備，愈來愈精良，目前市場上可以買到的各種廠牌的濃縮麥芽汁，品質都很好，尤其是使用罐裝的濃縮麥芽汁，讓剛開始的新手只要照著說明，按部就班的做，幾乎都能釀出不錯的啤酒。

要注意的是：一般市場或雜貨店裡賣的麥芽糖並不適合用來釀製啤酒，購買時要認清標籤，指名購買啤酒專用的濃縮麥芽汁。

＞＞＞濃縮麥芽汁介紹

市面上販賣的濃縮麥芽汁品質不盡相同，購買時要選擇百分之百純麥芽濃縮汁，有的品牌除了麥芽汁外，還添加了糖和其他原料，以下介紹最常用到的濃縮麥芽汁：

A.Light Malt Extract：顏色呈淡金黃的稻草色，釀出來的啤酒顏色很淺。

B.Amber Malt Extract：顏色呈漂亮的琥珀色，釀出來的啤酒有濃厚的口感和淡淡的焦糖香味。

C.Dark Malt Extract：顏色很深，呈現深褐色甚至黑褐色，釀製出來的啤酒帶有烘烤過的麥芽香。

D.Extra Light Malt Extract：是顏色最淡的濃縮麥芽汁，用來釀製顏色淺淡的英式啤酒或美式啤酒。

E.Wheat Malt Extract：小麥濃縮麥芽汁，可以釀出不同風味的小麥啤酒。

如何選購濃縮麥芽汁？

1.製造日期：濃縮麥芽汁因為含有極高濃度的麥芽糖，具有天然防腐的功能，不易變質，但是超過5年的陳年濃縮麥芽汁，品質仍會有少許的改變，顏色會變得比較深，麥芽的芳香會減退。

2.份量：濃縮麥芽汁分小大兩種包裝，小罐的重量在1.6～2公斤之間，大罐的重量是3公斤左右。

3.含糖量：濃縮麥芽汁的包裝上會注明加水稀釋後的含糖量，啤酒的酒精度大約是含糖量的一半，也就是說罐頭上注明的含糖量是9度（百分之9），啤酒大約是4.5%的酒精度，包裝上的說明要看清楚，有時它還要在釀製過程中添加額外的糖才能達到足夠的含糖量。

4.苦味含量：濃縮麥芽汁的苦味含量（Homebrewing Bitterness units）有HBU標記的濃縮麥芽汁表示添加了啤酒花，並且經過長時間的煮沸，啤酒花的α酸及β酸，已經被萃取到麥芽汁裡。

5.原料成份：大麥是濃縮麥芽汁的主要成份，不同的品牌有各種獨特的添加物，例如小麥、玉米、米等不同的穀類，甚至有的廠牌添加了焦糖，加深色澤；添加食用甘油，增加啤酒的質感；添加果汁或是香草創造出不一樣口味的啤酒。購買時要仔細了解罐頭內裝的是什麼原料，找出自己喜歡的濃縮麥芽汁，或嘗試一罐全新風味的驚奇！

1. Barley malt extract (comes in liquid form)

Canned malt extract is extracted from wort and is condensed into a syrup-like solution. It is preferred by beginners.

Factories that produce malt extract use traditional barley mashing methods. Use hot water to aggregate the malt sugar in the barley, then vacuum package it at low temperatures. The water evaporates leaving only 10% of its original volume in the thick malt sugar liquid. Because the equipment for vacuum packaging at low temperature is improving, the different brands of malt extract available all have solid quality, especially canned malt extract.

> > > Malt Extract Introduction

The malt extract sold in market comes in different varieties. Choose 100% pure malt extract to brew beer. Some brands contain not only malt extract, but additional sugar or other ingredients. The following are the more commonly used malt extracts:

i. Light Malt Extract: has light golden straw color giving the finished beer a light color.

ii. Amber Malt Extract: has a beautiful amber color giving the finished beer a thick tasty flavor and a light caramel aroma.

iii. Dark Malt Extract: dark brown or brownish black, the finished beer has a baked barley aroma.

iv. Extra Light Malt Extract: the lightest of all, it is used in brewing light colored English beers or American beers.

v. Wheat Malt Extract: extracted from wheat malt, it can be used to brew many different flavors of wheat beer.

Beginners just follow the instructions, step by step. Almost everyone can brew an acceptable beer.Note that maltose not suitable for brewing beer. Purchase malt extract specifically for brewing beer.

How to select malt extract:

1. The date : The malt extract contains very thick malt sugar and is naturally decay retardant. The quality will not easily decline, however, malt extract of five years of age may be of lower quality. The color will be darker and the aroma of the barley will dissipate.

2. Portion : It comes in two different sizes, large and small. A small can weighs between 1.6 ~ 2 kilos, while a large one is about 3 kilos.

3. Sugar content : On the package of the malt extract, the sugar content after the water is added will be clearly stated. The alcohol gravity of beer is about half the sugar content. If the can says that the sugar gravity is 9 (9%), the alcohol gravity will be about 4.5%. Read the instructions carefully. Sometimes a little extra sugar has to be added during the brewing process to reach the right sugar content.

4. Bitterness units : If the label says HBU (Homebrewing Bitterness Units) the malt extract is added with hops. After a long period of boiling, the α-acid and β-acid will be broken down in the wort.

5. The Ingredient : Barley is the main ingredient of malt extract. Different brands of extract have special additives, such as wheat, corn or rice. Some brands even have caramel added to make the color darker. Some add edible glycerin to increase the body of the beer, or juices or vanilla to vary the flavor. When purchasing, read the label carefully to understand what the ingredients are and to learn which is your favorite. Or sometimes you can try a completely new can of malt extract. The flavor might be a surprise.

2.濃縮麥芽粉

除了濃縮麥芽汁外，還有一種濃縮麥芽粉，也可以按照比例加水溶解，做為理想的釀啤酒原料，但濃縮麥芽粉的缺點是必需保持絕對的乾燥，它很容易吸收空氣中的水份，保存不妥當時，會變成像水泥塊似的不堪使用。當我釀啤酒時，除了濃縮麥芽汁外，我會再加入一部份的濃縮麥芽粉，增加啤酒多層次的香味口感及質感。

2. Dry Malt Extract (comes in powdered form)

It can be dissolved with a portion of water added, forming an ideal brewing ingredient. The downside is that it has to be kept absolutely dry, or it will absorb water from the air easily. If it is not stored well, it will harden like a rock.

When I brew beer, I always add some dry malt extract to increase the flavor and quality.

3.啤酒花

直到16世紀，啤酒業者才開始將啤酒花添加到麥芽汁中一起煮沸，增加啤酒的味道；啤酒花有防腐作用，可以增加啤酒保存的時間。

依《本草綱目》記載：啤酒花有健胃、利尿、鎮靜等功效。啤酒花是攀藤性植物，又名蛇麻草，雌性花朵才是啤酒花，層層花瓣緊緊包裹，有點像松樹結的果實松塔，花朵成熟時，底部會有顆粒狀的物質，它是賦予啤酒苦味及香氣的關鍵。啤酒花可以幫助蛋白質凝聚，澄清麥汁，提高啤酒的穩定性。

啤酒花適於生長在冷涼的寒帶，德國及捷克生產優質的啤酒花有很悠久的歷史，日本北部及中國東北也有生產；因品種及種植地區的不同，風味亦不完全相同，技巧的運用不同品種的啤酒花可以創造出啤酒濃烈或淺淡的苦味芳香或中性的口味，千變萬化讓人著迷。

市售的啤酒花有三種型式：

1. **整朵乾燥的啤酒花**：花朵在8～9月間尚未授粉前，整朵摘下來立刻乾燥處理，加入麥芽汁中煮沸時需要長時間的沸滾，花朵互相撞擊，產生同性異位化的原子變化，才會充份的釋放出啤酒花脂中含有的α及β酸，賦予啤酒爽口的苦味。
2. **塊狀啤酒花**：整朵乾燥的啤酒花，壓成塊狀加入麥芽汁中煮沸時就會散開，需要長時間的煮沸才能萃取出它的風味。
3. **顆粒狀**：乾燥的啤酒花，研磨成粉末再壓製成顆粒，因為研磨時已經破壞了啤酒花中的化學結構，加入麥芽汁中煮沸只要5～10分鐘，就可以釋放出啤酒花的α及β酸，是近代科學文明的產品；缺點是不易將它自麥芽汁中分離。家庭釀製啤酒採用顆粒狀啤酒花可以節省很多時間，分離時用咖啡濾紙有很好的效果。

3. Hops

Until the 16th century, the beer brewer began to add hops to the wort during the boiling to enhance the flavor of the beer. Hops has retards decay and prolongs the shelf life of the beer.

>>> 常用到啤酒花及其特色

1. **East Kent Goldings**：英式淡啤酒主要使用這種啤酒花，有精緻的苦味及優雅的芳香。
2. **Fuggles**：傳統的淡啤酒使用，不似East Kent Goldings幽雅，苦味及芳香比較淡，美國式的淡啤酒都用這種啤酒花。
3. **Cascade**：美國式Pale Ale的主要原料，苦味較淡，偏重於香味。
4. **Saaz**：捷克生產的王牌啤酒花，高雅、芳香，是頂級啤酒的催生者。
5. **Hallertau**：德式啤酒Lager的主要原料，清爽芳香。
6. **Eroica**：濃郁的苦澀及芳香，是黑啤酒不可缺少的原料。
7. **Northern Brewer**：適度的苦澀，怡人的芳香，淡啤酒、黑啤酒都能勝任是全能性的啤酒花。

According to "Ancient Medicinal Herbs", hops has the function of strengthening stomach, stabilizing urination and calming the heart. Hops are a climbing vine. The male flower, surrounded by layers of petals, forms the hops. It is like acorns from the pine tree. When the flowers blossom, a crumbly material forms at the base. It is the key to the bitterness and aroma of the beer. Hops helps the proteins to agglutinate to clarify the wort and increase the stability of the beer.

Hops is better grown in a cold climate. Germany and Czechoslovakia have a long history of producing high quality hops. Hops also grows in northern of Japan and in north-east of China. Due to different local varieties and characteristics, the flavor is not uniform. Skillful use of the different brands of hops can create varying textures and flavors with different levels of bitterness. Its endless variety is quite fascinating.

Commercially available hops come in three different types:

1. **Dried cone-like flowers:** The whole flowers are removed before producing pollen and dried immediately. When added to the wort, a longer period of boiling is needed. The flowers are crush against one another in the boiling liquid, undergoing atoms changes and releasing the α-acid and β-acid .
2. **Pieces:** The whole dried flowers are pressed into pieces. When added to the boiling wort they will separate. A longer boiling time is needed to break down its flavor.
3. **Pellets:** Dried flowers are ground into powder and pressed into pellets. Because the grinding process has already destroyed the chemical structures of the hops, when added to the boiling wort it will only take 5 ~ 10 minutes to release the α-acid and β-acid in hops. Pelletization is the product of recent scientific advances. The disadvantage is that it is not easy to separate it from the wort. Using this type of hops to brew beer at home will save much time. Coffee filters are a good way to separate it from the wort.

> > > Commonly used hops and their characteristics

1. **East Kent Goldings:** English light beer uses this hops, it has a subtle bitterness and a graceful aroma.
2. **Fuggles:** Used in traditional light beer, it is not like as graceful as East Kent Goldings, the bitterness and aroma are lighter. American light beers all use this type of hops.
3. **Cascade:** The main ingredient in American Pale Ale, the bitterness is lighter, while its essence lies in its aroma.
4. **Saaz:** The king of hops in Czechoslovakia, it is elegant and fragrant .
5. **Hallertau:** The main ingredient of German Lager, it is clear, light and fragrant.
6. **Eroica:** It has thick bitterness and aroma, it is a necessary ingredient in dark beer
7. **Northern Brewer:** The bitterness is quite appropriate with a pleasant aroma. It is a type of hops with many applications in both light and dark beers.

4.酵母菌

市售啤酒的種類很多，尤其是歐美各國的超市，各種各樣的名稱真會讓你不知如何挑選，其實啤酒基本上只有兩種，英式啤酒Ale是利用上層發酵的酵母菌釀製而成，德式啤酒Lager是利用下層發酵的酵母菌釀製而成。

以下將Ale啤酒及Lager啤酒使用的酵母菌不同之處做一個比較：

	Ale酵母菌	Lager酵母菌
發酵溫度	16～18˚C（60～65˚F），發酵需要較高的溫度。	7～10˚C （45～50˚F），發酵溫度低。
發酵時間	發酵作用快速，約一週即可完成。	發酵作用慢，需3～5週才會完成。
發酵時	酵母浮在表面	酵母沉在底部
香氣	產生水果芳香味，如香蕉、鳳梨等風味。	沒有特殊香味，但是低溫長時間發酵使得啤酒有清新爽口的特色。
發酵完成時	酵母沉澱到底部	酵母沉澱到底部

■啤酒酵母菌有冷凍乾燥製成粉末狀的酵母菌和液體活菌酵母菌，台灣目前只有粉末狀的乾燥酵母菌，使用前要將粉狀酵母放入一杯溫水（38˚C）中，等候30分鐘後，讓脫水的酵母在溫水中活化，再加入麥芽汁中進行發酵作用。

4. Yeast

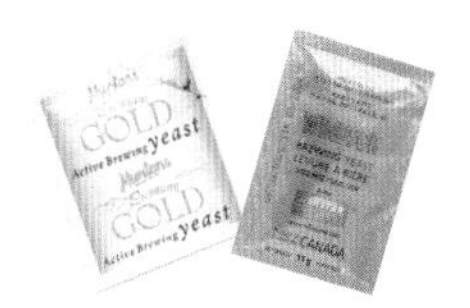

There are many yeast brands commercially available, especially in European and American supermarkets. With all the different names it can be difficult to pick the right one. In fact there are only two kinds of brewer's yeast, English Ale, which is a top-fermenting yeast and German Lager, a bottom-fermenting yeast.

The table below shows the differences between Ale and Lager:

	Ale Yeast	Lager
Fermentation Temperature	16~18C (65~65F), needs warmth to ferment	7~10C (45~50F), needs a cool temperature to ferment
Fermentation Time	Fermentation goes rapidly, and takes about one week	Fermentation is slow, taking about 3~5 weeks
During Fermentation	The yeast floats on the surface	The yeast sinks to the bottom
Aroma	Fermentation produces a fruity aroma, resembling banana or pineapple	There is no special aroma, however, the cool temperature and long period of fermenting make the beer light and clear
Fermentation Finished	The yeast sinks to the bottom	The yeast sinks to the bottom

Yeast is available in dried powdered and liquid forms. Only dried powdered yeast is currently sold in Taiwan. Dissolve the powdered yeast in a cup of warm water (38C) to let the yeast revive in the warm water, then pitch into the wort to begin fermentation.

5. 果糖

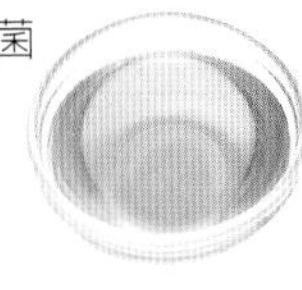

在新釀啤酒第二次發酵時加入少許果糖，讓酵母菌重新獲得營養補充，恢復活力，酵母菌就又開始了它天生的本能，進行發酵作用，產生酒精及二氧化碳，才會有啤酒特有的氣泡。

5. Fructose

Add a little fructose during the secondary fermentation to let the yeast regain its nutrient supplement and recover its energy. Yeast will start its natural capability and carry on the fermentation once again, then alcohol and CO2 will be produced and form the special beer foams.

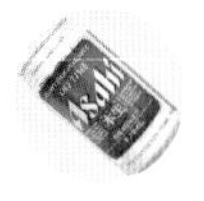

6.吉利丁

吉利丁使用方便，是家庭釀製啤酒最常使用的澄清劑。

吉利丁使用方法：

份量：20公升的麥芽汁需使用5公克的吉利丁（購買素食用的吉利丁，否則會有異味）

（1） 吉利丁溶解於1杯冷水中，等候10分鐘。

（2） 將溶入冷水的吉利丁，放入雙層鍋中加熱至約70℃，呈透明狀即可；千萬不能煮開，吉利丁煮沸後就失去了澄清的作用。

（3） 倒入麥芽汁中，攪拌均勻。

6. Gelatin

Gelatin is convenient to use and is the most common clarifying agent in home brewing.

How to use gelatin:

Portion: 20 liters of wort needs about 5 grams of gelatin (purchase vegetarian gelatin, or it will taste strange.)

(1) Dissolve the gelatin in 1 cup of cold water and let sit for 10 minutes.

(2) Remove the cold water to a double-boiler and heat up to 70C until transparent. Do not bring to a boil, or the gelatin will lose its ability to clarify.

(3) Pour into the wort and stir to mix.

7.水

釀製啤酒需要大量的水，含有礦物質的硬水不適合釀製啤酒，城市裡的自來水是最方便取得的水，但自來水含有氯，使用前要把自來水煮沸讓水中含有的氯揮發。目前很流行的RO逆滲透水，經過多重過濾，已經去除了大部份的礦物質，用來釀啤酒很理想。

7. Water

Brewing beer requires large amounts of water. Hard water which contains minerals is not suitable for brewing. Tap water is the most convenient water for brewing, but as it is often fluoridated, boil the tap water to let the fluoride evaporate before using it. Water that has been filtered many times, resulting in completely removal of minerals, is ideal for brewing.

STEP BY STEP輕鬆釀啤酒

使用濃縮麥芽汁，最快只要2週就可以喝到自釀啤酒，很適合現代人求新求快的心理，最主要還是因為罐裝濃縮麥芽汁的種類多，品質穩定，可以千變萬化的釀製出各種不同風味的英式啤酒Ale及德式啤酒Lager。

我們先示範一個最基本型的，以濃縮麥芽汁釀製的英式淡啤酒（English Ale），帶有淡淡的苦味及精緻的啤酒花香，酒精度不高，發酵一個星期就可以裝瓶，兩週就可以暢飲；不必苦苦等待，很快就可以驗收成果。使用罐裝麥芽汁的便利之處是罐上均注明了糖的比重含量，只要按著說明添加水量，就不必擔心稀釋後的麥芽汁含糖量不足。

Brew Beer with Ease (STEP BY STEP)

IV. Use Malt Extract Liquid and Malt Extract Powder to Brew Beer at Home with Ease (STEP BY STEP)

Using malt extract will enable you to enjoy your homebrewed beer within as little as 2 weeks. It is perfect for us moderns, who want things to be fast and new. Most importantly, using canned malt extract liquid enables brewing of many types of beer with stable quality, such as English Ale and German Lager

Here we would like to demonstrate a basic type beer, which is brewed with malt extract liquid - English Ale. It has a light bitterness and fine flowery aroma with low alcohol gravity. After fermentation, it can be bottled in a week and enjoyed in 2 weeks. You can taste the results in no time. One good thing about using canned malt extract liquid is that the label clearly states the sugar content. If you follow the instructions and add water, you do not have to worry about whether the malt liquid has sufficient sugar.

英式淡啤酒 English Ale

麥芽汁

麥芽粉

啤酒花顆粒

啤酒酵母菌

果糖

吉利丁

材料：

Extra Light Malt Extract濃縮麥芽汁2公斤
英式啤酒Ale濃縮麥芽粉1公斤
啤酒花15公克
英式啤酒Ale酵母菌1包
果糖3/4杯（180 ml.）
素食用吉利丁片5公克

Ingredients:

2kg extra light malt extract
1kg English Ale malt extract powder
15g hops
1 pack English Ale yeast
3/4C (180ml.) fructose
5g vegetarian gelatin pieces

英式淡啤酒 English Ale

請各位和我一同來溫習一遍啤酒釀製的過程：

首先，最重要得是：清洗乾淨所有的設備，以偏亞硫酸鉀水溶液消毒。這個步驟是釀酒能成功最重要的關鍵，好的開始是成功的一半，良好的衛生習慣保證會帶給你好的結果，髒亂的環境及設備不但會污染了使用的原料，而且危害了自己的健康，偏亞硫酸鉀味道稍嗆，工作時請戴口罩。

取一個洗淨的塑膠桶，放入4公升的水及60公克的偏亞硫酸鉀，充份溶解後把要用的工具，如長柄匙、發酵鎖、虹吸管、濾網、比重測糖計等，全部泡在裡面，當你要用這些工具時再從裡面拿出來。

Let us review the procedures of brewing beer one more time:

First, the most important thing is: clean all the equipment thoroughly and sterilize with potassium metabisulfite.

This procedure is the key to a successful beer. Half begun is well done. Good hygiene habits will bring good results. A dirty, sloppy environment will contaminate all the ingredients and endanger your own health. Potassium metabisulfite is toxic, so please wear masks when working.

Combine 4 liters of water and 60 grams of potassium metabisulfite in a clean plastic bucket thoroughly. Soak all the utensils you will use in the water, from the long handled stirring paddle to the airlocks, siphon tubes, sieve and hydrometer. Remove when they are needed.

步驟圖Procedure Chart

1. 整罐的麥芽汁放入一盆熱水中浸泡。（罐中的麥芽汁會軟化，開罐後容易倒出。）

1. Soak the whole can of malt extract liquid in a basin of hot water. (To help the liquid soften, so that it is easier to pour out when it is opened.)

2. 不銹鋼大鍋中倒入6公升的水煮開，熄火。

2. Bring 6 liters of water in a stainless steel pot to a vigorous boil, then remove from heat.

3. 倒入整罐麥芽汁，再舀出一些鍋中開水淌出罐底及四周不易倒出的麥芽汁。

3. Pour the whole can of malt extract liquid into the water, then scoop some water from the pot to rinse the malt liquid stuck around the sides of the can.

4.以長柄匙充份攪拌溶解。

4. Stir vigorously with a stirring paddle.

5.加入麥芽濃縮粉，攪拌溶解。

5. Add malt extract powder, then stir until dissolved completely.

6.開火，再次煮沸麥芽汁15分鐘，加入15公克啤酒花顆粒。

6. Return to heat and boil the solution for 15 minutes, then pitch in 15 grams of hops.

7.繼續煮5分鐘。

7. Continue boiling for 5 more minutes.

8.就製成了麥芽汁wort。

8. The wort is prepared.

9.將6公升的冰開水裝入25公升的玻璃發酵瓶中。

9. Pour 6 liters of ice cold water inside the 25-liter glass fermenter.

10.以濾網將wort小心的過濾出雜質，倒入玻璃發酵瓶中。

10. Pour the wort through a sieve to remove any impurities and dregs.

11.將濾網中的啤酒花含有的汁液輕輕壓擠滴入發酵瓶中。

11. Squeeze the liquid out of the hops through a sieve gently into the fermenter.

12.取出少許的麥芽汁，降溫到16℃，用比重測糖計測出糖的含量，記錄在筆記本上。

12. Remove a little of wort and lower the temperature to 16C. Measure the sugar gravity with a hydrometer and make a record.

13.加入冷水直到發酵瓶中有19公升的麥芽汁。

13. Add cold water until the fermenter contains 19 liters of wort.

14.發酵桶中的麥芽汁降溫到24°C時，加入啤酒酵母。

14. When the temperature of the wort in the fermenter has fallen to 24C, pitch in the yeast.

15.粉末狀的乾燥酵母菌使用前要將粉狀酵母放入一杯溫水（38°C）中。

15. Before being added, the dried yeast powder has to be dissolved in a cup of warm water (38C).

16.等候30分鐘後，讓脫水的酵母在溫水中活化。

16. Let sit for 30 minutes to let the yeast activate in the warm water.

17.將酵母加入麥芽汁中進行發酵作用，用長柄匙攪拌均勻。

17. Pitch in the wort and start the fermentation process, stir vigorously until evenly mixed.

18.發酵瓶加蓋，並加上發酵鎖，將發酵瓶放置在（18～21°C）的陰暗處。

18. Put the lid of the fermenter on and place the airlocks in the lid. Remove the fermenter to a dark cool place (18~21C).

19.24小時後檢查發酵作用是否開始，發酵鎖有氣泡不斷的湧出表示發酵正常。

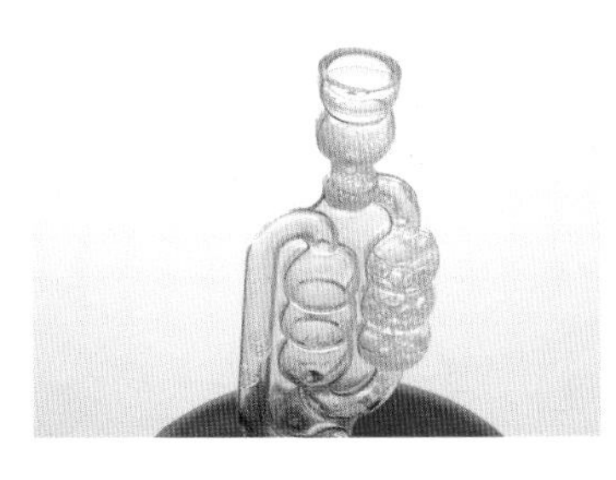

19. 24 hours later, check if the fermentation has started, if foams keep pushing the airlock out of the lid, the fermentation is going on well.

20.英式啤酒Ale使用的上層發酵作用酵母菌大約在一週內就會經歷分裂繁殖發酵，沈澱不同的過程，完成它的生命週期。當發酵鎖湧出的氣泡逐漸減弱終於停止，此時再倒出一些麥芽汁以比重測糖計檢測含糖量，應該已經降到0度左右，就可以進行裝瓶前的準備工作。

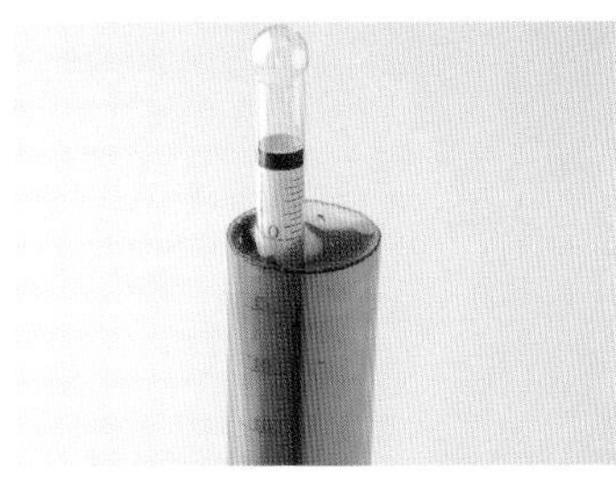

20. English Ale uses top-fermenting yeast which takes about 1 week to exhaust the sugars in the wort and sinks to the bottom and finish its life circle. When the foam stops pushing out the airlock, remove a little wort and measure the sugar gravity with the hydrometer, it should be down to 0. if so, it is time to rack to bottles.

裝瓶前的準備工作／Preparation before Bottling

新釀啤酒只是經過發酵作用，而含有酒精的麥芽汁，並不像我們習慣喝的啤酒，清涼有勁充滿了躍動的氣泡。要經過第二次發酵處理、裝瓶後，才會產生出啤酒的特色及風味，這也就是所謂的裝瓶前的準備工作。

The finished beer is wort with alcohol, which just went through the primary fermentation, unlike the light cool beer full with a full head of foam that we are used to drinking. After the secondary fermentation and bottling, the characteristics and flavors of beer will be produced. Prepare as follows:

步驟圖Procedure Chart

1.清洗所有要用的啤酒瓶，用高溫熱水或是偏亞硫酸鉀水溶液消毒（60公克的偏亞硫酸鉀，溶解在4公升的冷水中）

1. Rinse all the beer bottles, use high temperature hot water or potassium metabisulfite to sterilize (dissolve 60 grams of potassium metabisulfite in 4 liters of cold water).

2.第一個25公升的玻璃瓶已經有19公升發酵完成的新釀啤酒，第二個25公升的玻璃瓶也清洗消毒，同時清洗消毒等一下要用的啤酒蓋、虹吸管、長柄匙及啤酒瓶，浸泡在偏亞硫酸鉀水溶液中30分鐘。

2. The first 25-liter glass fermenter already holds 19 liters of fermented finished beer. The second 25-liter fermenter has to be sterilized, along with the bottle caps, siphon tube, long handledstirring paddle and beer bottles, in potassium metabisulfite solution for 30 minutes.

3.將裝滿新釀啤酒的玻璃瓶移放到較高的位置，例如桌上或椅子上；另外的空瓶放在它的下面較低的位置，利用虹吸管將新釀啤酒上層的澄清液吸到下面的空瓶裡。虹吸時要小心不要把上面發酵瓶底部的沈澱物吸出來。

3. Place the first glass fermenter with the finished beer at a higher position, for example, on a table or chair. Place the empty fermenter at a lower position (see photo), and use a siphon tube to convey the clear liquid at the top to the empty fermenter. When siphoning, please do not take up the sediment at the bottom.

4. 兩杯水（480ml.）加入3/4杯（180ml.）果糖，放入清潔的鍋中煮開。這份糖水要加到新釀啤酒中供給酵母菌營養，讓少量留存在新釀啤酒中的酵母菌恢復生機，再次的進行分裂繁殖發酵作用，產生二氧化碳氣，啤酒開瓶時才會產生氣泡。

4. Boil two cups of water (480ml.) with 3/4 cup of fructose (180ml.) in a sterilized pan. The syrup is for adding to the finished beer to provide nutrients for the remaining yeast and reactivate it for growth and reproduction. Carbonation will also be produced, so that the beer will foam when the bottle is opened.

5. 糖水倒入新釀啤酒中，用長柄匙攪拌均勻。

5. Add syrup to the finished beer and stir with the stirring paddle.

6. 將5公克吉利丁粉溶解於300 ml.的冷水中，10分鐘後，隔水加熱成透明狀，但不能煮沸，吉利丁液加入新釀啤酒中，讓懸浮物沈降，啤酒才會清澈，如果讓吉利丁液煮沸，就失去了澄清的作用。

6. Dissolve 5 grams of gelatin powder in 300ml. of cold water. Let sit for 10 minutes. Heat in a double-boiler until transparent, but not boiling. Add gelatin solution to the finished beer, so that the floating objects can sink to the bottom and the beer will be clear and clean. If the gelatin solution is boiling, it will lose its ability to clarify.

7. 吉利丁液倒入新釀啤酒中，用長柄匙攪拌均勻。

7. Add gelatin solution to the finished beer and stir with a stirring paddle.

8.將浸泡在偏亞硫酸鉀水溶液中的啤酒瓶拿出來，用熱水裡外沖洗一下，以免餘留的偏亞硫酸鉀抑制了酵母菌的作用。利用虹吸管，將第二次發酵處理完成的新釀啤酒裝滿啤酒瓶。

8. Remove the beer bottles, which should still be soaking in the potassium metabisulfite solution. Rinse with hot water to prevent the potassium metabisulfite from killing the yeast.Rack the finished beer to the beer bottles with a siphon tube.

9.不要裝的太滿，餘留大約2吋的空間（如圖所示），留的空間不夠，酵母菌沒有足夠的空氣進行發酵作用，不容易產生足夠的二氧化碳氣，如果餘留了太多的空間，啤酒瓶可能會受不了過多二氧化碳的壓力而爆裂。

9. Do not overfill, but leave about 2 inches of space (see photo). If the space is not enough for the yeast to grow, there will not be enough carbon dioxide. Contrariwise, the beer bottles might explode from the pressure caused by too much carbon dioxide.

10.用最簡單型的裝蓋器，將啤酒瓶加蓋密封。可貼上你自己設計的標籤，注明裝瓶日期及啤酒種類。

10. Use a simple utensil to cap and secure. Label to record the date and the type of beer.

11.剛裝瓶的啤酒要放在陰暗、陰涼的地方熟成一段時間，讓剛加進去的糖及酵母菌進行發酵作用。這樣才會產生足夠的二氧化碳氣，同時增加啤酒風味及口感，上層發酵釀製的英式啤酒Ales，最好放置1～2週再喝，下層發酵釀製的德式啤酒Lagers，則要放置3～5週後才會好喝。

11. Place the beer in the dark cool place to age for some time. The yeast will ferment the sugar and produce enough carbon dioxide to increase the beer flavor and aroma. English Ale with top-fermenting yeast is better aged for 1~ 2 weeks before drinking. German Lagers with bottom-fermenting yeast should age for about 3 ~ 5 weeks before drinking.

狀況分析

釀製啤酒時會發生的不良狀況

1. 味道不純正

A. 苦澀味：形成的原因為當麥芽浸熱水時溫度太高，溶出過多的單寧而增加了苦澀的風味。

B. 厚紙板味：發酵過程中，接觸太多空氣而使麥芽汁氧化。

C. 酸味：啤酒不該有酸味，出現酸味是在釀造過程中遭到醋酸菌的感染而變質。

D. 煮熟的蔬菜味：麥芽汁受到微生物（雜菌）污染，表示在釀製過程中沒有做好消毒殺菌的工作。

2. 混濁：麥芽中的澱粉沒有充份轉化成麥芽糖，形成混濁不清的啤酒。

3. 缺乏氣泡：啤酒瓶不乾淨，或是瓶蓋沒有蓋緊。

4. 沒有質感：啤酒喝起來清淡似水，可能是麥芽汁中含有過量的糖，並非純的麥芽汁。或是裝瓶後進行第二次發酵時溫度太低，酵母菌不能將糖轉化成酒精及二氧化碳。

5. 瓶頸有一圈殘留物：表示瓶內的啤酒已經遭到污染，所以釀製時的清潔殺菌必須徹底。

Troubleshooting

Problems that might occur during brewing

1 . The flavor is not right

A. Dry flavor: The temperature of the hot water for soaking the barley is too high, and too much tannin has dissolved in the water, giving the brew a dry flavor.

B. Cardboard flavor: Oxidization of the wort during fermentation due to too much contact with the air.

C. Sour taste: Beer should not be sour. Sour beer results when the beer is contaminated by bacteria during the brewing process.

D.Cooked Vegetables: The wort has contaminated have to improve the sanitization & sterilization.

2 . Cloudy Liquid: The starch in the barley has not completely broken down into malt sugar, making the beer cloudy.

3 . Not enough foam: The beer bottles are not clean or the cap is not secure.

4 . Low beer quality: The beer tastes like water. One reason is that the malt extract contains too much sugar and is not pure malt extract. Or the temperature is too low during the secondary fermentation, so that the yeast cannot convert sugar into alcohol and carbon dioxide.

5 . There is a round stain around the neck of the bottle: It means that the bottle has been contaminated. Sterilization must be thorough.

Beer Bre

Chapter II

市售的濃縮麥芽汁種類相當多，歐洲、美洲、加拿大、澳洲都有各具特色的產品，本書為你介紹最具代表性的濃縮麥芽汁釀製啤酒：

加拿大淡啤酒Canadian Ale
美國式淡啤酒American Golden Ale
英國南方風味棕色啤酒Southern English Brown Ale
德國式啤酒German Lager
愛爾蘭苦啤酒Irish Stout
加州式低熱量淡啤酒California Style Low Calorie Light Beer
捷克式重口味啤酒Pilsner
墨西哥式啤酒Mexican
啤酒色香味的品評Beer tasting: the aroma, the color and the flavor.

ving

2

釀酒篇

Canadian Ale
加拿大淡啤酒

製成份量：約0.6公升×30瓶
需要時間：4週

材料：

Canadian Ale	
濃縮麥芽汁	1罐（2公斤）
Plain dry Malt Extract	
濃縮麥芽粉	1公斤
East Kent Goldings	
啤酒花顆粒	15公克
Ale啤酒酵母菌	1包
果糖	3/4杯（180ml.）
素食用吉利丁	5公克

加拿大及美國製作麥芽汁的大麥品種和歐洲使用的大麥不同，顆粒小風味比較淡。加拿大淡啤酒以較高溫度發酵，再低溫冷藏熟成的方式釀製。

The barley strain that Canadian and American ales use to prepare the wort is different than the barley that European ales use. The barley has small pearls and the flavor is lighter. Canadian light beer is fermented at a high temperature and aged at a low temperature.

Ingredients

1 can (2kg) Canadian Ale malt extract liquid
1kg plain dry malt extract
15g East Kent Goldings hops pellets
1 pack Ale yeast
3/4C (180ml.) fructose
5g vegetarian gelatin

Portion Preparation: approximately 0.6 liters x 30 bottles
Time Needed: 4 weeks

麥芽汁

麥芽粉

啤酒花顆粒

啤酒酵母菌

果糖

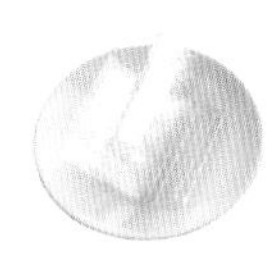
吉利丁

Canadian Ale

前製作業

取一洗淨的塑膠桶，放入4公升的水及60公克的偏亞硫酸鉀，充份溶解後把要用的工具，如長柄匙、發酵鎖、虹吸管、濾網、比重測糖計等，全部泡在裡面。

做法

1.不銹鋼大鍋中倒入6公升的水煮開，加入Canadian Ale濃縮麥芽汁煮沸15分鐘，熄火前放入East Kent Goldings啤酒花顆粒，製成麥芽汁。
2.將10公升冰開水倒入玻璃發酵瓶中，將麥芽汁過濾倒入。
3.添加冷水，直到瓶中有19公升的麥芽汁。加入Plain dry Malt Extract濃縮麥芽粉，攪拌溶解。
4.取出半杯麥芽汁，放冷後（約16°C）用比重測糖計量取糖的含量，紀錄在筆記本。
5.當麥芽汁降溫到24°C時，加入Ale酵母菌，以長柄匙攪拌均勻。
6.放置在20°C左右的陰暗處發酵。
7.24小時後檢查發酵作用是否開始，發酵鎖有氣泡不斷的湧出表示發酵正常。
8.約一週後，當發酵鎖湧出的氣泡逐漸減弱而停止時，再倒出一些麥芽汁以比重測糖計檢測含糖量，降到0度左右時即表示第一次發酵作用完成，虹吸換瓶。

Preparation

In a clean plastic bucket combine 4 liters of water with 60 grams of potassium metabisulfite, stir until dissolved completely then soak all the utensils, such as the long handled stirring paddle, airlocks, siphon tubes, sieve and hydrometer, in the solution.

Methods:

1. To prepare wort: Boil 6 liters of water in a stainless steel pan, add Canadian Ale malt extract liquid. Bring to a vigorous boil for 15 minutes, then add East Kent Goldings hops pellets before removing from heat.
2. Fill 25 liter glass fermeater with 10 liters of ice-cold water. Pour wort into it through a sieve to remove any dregs.
3. Add cold water as needed until the fermenter holds 19 liters of liquid. Add plain dry malt extract, stir until well-dissolved.
4. Remove half a cup of wort and let cool (around 16C). Measure with a hydrometer to take the sugar gravity and record.
5. When the temperature of the wort is down to 24C, pitch in the yeast and stir with a stirring paddle until well-mixed.

6. Let ferment in dark place at around 20C.
7. Wait for 24 hours and check if the fermentation has started. If the foam keeps bubbling out the airlocks, it shows that fermentation is going on normally.
8. Let ferment for about 1 week. When the foam stops bubbling out the airlock, pour in a little wort and measure with a hydrometer to test the sugar gravity. If it is around 0, it means that the primary fermentation has finished, siphon and bottle.

第二次發酵及澄清處理

9. 將兩杯水與3/4杯果糖一起放入鍋中煮開。
10. 吉利丁溶解於300 ml.冷水中，隔水加熱成透明狀。
11. 糖水和吉利丁液倒入新釀啤酒中，以長柄匙攪拌均勻。
12. 將啤酒瓶浸泡在偏亞硫酸鉀水溶液消毒後，再用熱水裡外沖洗掉偏亞硫酸鉀水。
13. 利用虹吸管，將新釀啤酒裝滿啤酒瓶，不要裝的太滿，餘留大約2吋的空間。加蓋密封。
14. 放置在6°C左右陰暗的地方，熟成2週。

Secondary fermentation and clarification

9. Boil two cups of water with 3/4 cup of fructose in pan.
10. Dissolve gelatin in 300ml. of cold water, heat in a double-boiler until transparent.
11. Pour the sugar solution and gelatin into the finished beer, stir with a stirring paddle until evenly mixed.
12. Rinse the empty beer bottles thoroughly to get rid of the potassium metabisulfite solution after removing from the potassium metabisulfite solution for sterilizing.
13. Siphon with a siphon tube and fill the empty beer bottles with the finished beer. Do not fill up all the way, leave about 2 inches of space, then seal with a cap.
14. Let age in a cool dark place at about 6C for 2 weeks.

Americen Golden Ale

美國式淡啤酒

- 製成份量：約0.6公升×30瓶
- 需要時間：3～4週

和加拿大淡啤酒像是雙生兄弟，使用的原料不一樣，採用相同的釀製程序，經過低溫熟成口感清爽，是炎炎夏日的熱門飲料。

Californian and Canadian light beer are twins, the ingredients are different, but the brewing procedures are the same. After low temperature aging, the flavor is light and clear. It is a popular drink on hot summer days.

材料：

Pale Ale濃縮麥芽汁1罐	2公斤
extra light濃縮麥芽粉	1公斤
Cascade啤酒花顆粒	10公克
Ale啤酒酵母菌	1包
果糖	3/4杯（180ml.）
素食用吉利丁	5公克

Ingredients

1 can (2kg) Pale Ale malt extract liquid
1kg extra light dry malt extract
10g Cascade hops pellets
1 pack Ale yeast
3/4C (180ml.) fructose
5g vegetarian gelatin

- Portion Preparation: approximately 0.6 liters x 30 bottles
- Time Needed: 3 ~ 4 weeks

麥芽汁

麥芽粉

啤酒花顆粒

啤酒酵母菌

果糖

吉利丁

California Golden Ale

前製作業

取一洗淨的塑膠桶，放入4公升的水及60公克的偏亞硫酸鉀，充份溶解後把要用的工具，如長柄匙、發酵鎖、虹吸管、濾網、比重測糖計等，全部泡在裡面。

做法

1 .不銹鋼大鍋中倒入6公升的水煮開，加入Pale Ale濃縮麥芽汁煮沸15分鐘，熄火前放入Cascade啤酒花顆粒，製成麥芽汁。

2 .將10公升冰開水倒入玻璃發酵瓶中，將麥芽汁過濾倒入。

3 .添加冷水，直到瓶中有19公升的麥芽汁。加入extra light濃縮麥芽粉，攪拌溶解。

4 .取出半杯麥芽汁，放冷後（約16℃）用比重測糖計量取糖的含量，紀錄在筆記本。

5 .當麥芽汁降溫到24℃時，加入Ale酵母菌，以長柄匙攪拌均勻。

6 .放置在20℃左右的陰暗處發酵。

7 .24小時後檢查發酵作用是否開始，發酵鎖有氣泡不斷的湧出表示發酵正常。

8 .約一週後，當發酵鎖湧出的氣泡逐漸減弱而停止時，再倒出一些麥芽汁以比重測糖計檢測含糖量，降到0度左右時即表示第一次發酵作用完成，虹吸換瓶。

Preparation

In a clean plastic bucket combine 4 liters of water with 60 grams of potassium metabisulfite, stir until dissolved completel then soak all the necessary utensils, such as the long handled stirring paddle, airlocks, siphon tubes, sieve and hydrometer, i the solution.

Methods:

1. To prepare wort: Boil 6 liters of water in a stainless steel pan, add Pale Ale malt extract liquid. Let boil vigorously for 1 minutes, then add Cascade hops pellets before removing from heat.
2. Fill 25 liter glass fermeater with 10 liters of ice-cold water. Pour wort into it through a sieve to remove any dregs.
3. Add cold water as needed until the fermenter holds 19 liters of liquid. Add extra light dry malt extract, stir until well-di solved.
4. Remove half cup of wort and let cool (around 16C). Measure with a hydrometer to test the sugar gravity and record.
5. When the temperature of the wort is down to 24C, pitch in the yeast and stir with a stirring paddle until well-mixed.

6. Let ferment in a dark place at around 20C.
7. Wait for 24 hours and check if the fermentation has started. If bubbles keep foaming out of the airlocks, it shows that the fermentation is going on normally.
8. Let ferment for about 1 week. When the bubbles finally stop foaming out the airlocks, pour a out little wort and measure with a hydrometer to test the sugar gravity. If it is around 0, it means that the primary fermentation has finished. Siphon and bottle.

第二次發酵及澄清處理

9. 將兩杯水與3/4杯果糖一起放入鍋中煮開。
10. 吉利丁溶解於300 ml.冷水中，隔水加熱成透明狀。
11. 糖水和吉利丁液倒入新釀啤酒中，以長柄匙攪拌均勻。
12. 將啤酒瓶浸泡在偏亞硫酸鉀水溶液消毒後，再用熱水裡外沖洗掉偏亞硫酸鉀水。
13. 利用虹吸管，將新釀啤酒裝滿啤酒瓶，不要裝的太滿，餘留大約2吋的空間。加蓋密封。
14. 放置在6°C左右陰暗的地方，熟成2週。

Secondary fermentation and clarification

9. Boil two cups of water with 3/4 cup of fructose in pan.
10. Dissolve gelatin in 300ml. of cold water, heat in a double-boiler until transparent.
11. Pour the sugar solution and gelatin into the finished beer, stir with a stirring paddle until evenly mixed.
12. Rinse the empty beer bottles thoroughly to get rid of the potassium metabisulfite solution after removing from the potassium metabisulfite solution for sterilizing.
13. Siphon with a siphon tube and fill the empty beer bottles with the finished beer. Do not fill up all the way, leave about 2 inches of space, then seal with a cap.
14. Let age in a cool dark place at about 6C for 2 weeks.

Brown Ale
英國南方風味棕色啤酒

製成份量：約0.6公升×30瓶
需要時間：3週半

英國南方工業城市Newcastle最早釀製出此種啤酒而帶動風潮，口味很柔順，酒精度不高色澤呈棕紅色屬於Dark Ale系列啤酒。

The southern England city of Newcastle is an industrial town. Newcastle was the first place this beer was brewed and introduced to the world. The flavor is smooth, and it has a with low alcohol gravity. A dark ale, the color is reddish brown.

材料：

Brown Ale濃縮麥芽汁	1罐（2公斤）
Chocolate Malt	500公克
Plain dry濃縮麥芽粉	1公斤
Fuggles啤酒花顆粒	15公克（苦味）
Fuggles啤酒花顆粒	10公克（芳香）
Ale啤酒酵母菌	1包
果糖	1/2杯（125ml.）
素食用吉利丁	5公克

- Chocolate Malt為大麥浸水發芽後烘焙時，以高溫烘焙成巧克力的顏色，主要作用是增加啤酒的色澤。
- 第一次加入的啤酒花是要萃取啤酒花的苦味，關火前或關火後再加入的啤酒花，則為了萃取啤酒花的香味。

Ingredients

1 can (2kg) Brown Ale malt extract liquid
500g Chocolate malt
1kg plain dry malt extract
15g Fuggles hops pellets (bitter agent)
10g Fuggles hops pellets (aroma)
1 pack Ale yeast
1/2C (125ml.) fructose
5g vegetarian gelatin

Portion Preparation: approximately 0.6 liters x 30 bottles
Time Needed: 3.5 weeks

* Chocolate malt is prepared when the soaked barley is roasted under high temperature, turning the color chocolate brown. Its function is to enhance the color of the beer.

* The first hops is for its bitterness, the second hops added just before or after removing from heat is included for its aroma.

麥芽汁

Chocolate Malt

麥芽粉

啤酒花顆粒

啤酒酵母菌

果糖

Brown Ale

前製作業

取一洗淨的塑膠桶，放入4公升的水及60公克的偏亞硫酸鉀，充份溶解後把要用的工具，如長柄匙、發酵鎖、虹吸管、濾網、比重測糖計等，全部泡在裡面。

做法

1. Chocolate Malt壓碎，放入不銹鋼大鍋中，與6公升的水煮沸30分鐘。加入Brown Ale濃縮麥芽汁及Fuggles啤酒花15公克煮沸15分鐘，再放入Fuggles啤酒花10公克煮沸2分鐘熄火，製成麥芽汁。
2. 將10公升的冰開水倒入玻璃發酵瓶中，將麥芽汁過濾倒入。S
3. 添加冷水，直到瓶中有19公升的麥芽汁。加入Plain dry濃縮麥芽粉，攪拌溶解。
4. 取出半杯麥芽汁，放冷後（約16℃）用比重測糖計量取糖的含量，紀錄在筆記本。
5. 當麥芽汁降溫到24℃時，加入Ale酵母菌，以長柄匙攪拌均勻。
6. 放置在20℃左右的陰暗處發酵。
7. 24小時後檢查發酵作用是否開始，發酵鎖有氣泡不斷的湧出表示發酵正常。
8. 約10天，當發酵鎖湧出的氣泡逐漸減弱而停止時，再倒出一些麥芽汁以比重測糖計檢測含糖量，降到0度左右時即表示第一次發酵作用完成，虹吸換瓶。

Preparation

In a clean plastic bucket combine 4 liters of water with 60 grams of potassium metabisulfite, stir until dissolved completely then soak all the necessary utensils, such as the long handled stirring paddle, airlocks, siphon tubes, sieve and hydrometer, in the solution.

Methods:

1. To prepare wort: Crush chocolate malt and boil in 6 liters of water for 30 minutes. Add Brown Ale malt extract liquid and 15 grams of Fuggles hop crums. Bring to a rolling boil for 15 minutes, then add 10 grams of Fuggles hops pellets. Boil for 2 minutes and remove from heat.
2. Fill 25 liter glass fermeater with 10 liters of ice-cold water. Pour wort into it through a sieve to remove any dregs.
3. Add cold water as needed until the fermenter holds 19 liters of liquid. Add plain dry malt extract, stir until well-dissolved.
4. Remove half cup of wort and let cool (around 16C). Measure with a hydrometer to test the sugar gravity and record.

5. When the temperature of the wort is down to 24C, pitch in the yeast and stir with a stirring paddle until well-mixed.
6. Let ferment in a dark place at around 20C.
7. Wait for 24 hours and check if the fermentation has started. If bubbles keep foaming out of the airlocks, it shows that the fermentation is going on normally.
8. Let ferment for about 10 days. When the bubbles finally stop foaming out the airlocks, pour in a little wort and measure with a hydrometer to test the sugar gravity. If it is around 0, it means that the primary fermentation has finished. Siphon and bottle.

Brewing

第二次發酵及澄清處理

9. 將兩杯水與3/4杯果糖一起放入鍋中煮開。
10. 吉利丁溶解於300 ml.冷水中，隔水加熱成透明狀。
11. 糖水和吉利丁液倒入新釀啤酒中，以長柄匙攪拌均勻。
12. 將啤酒瓶浸泡在偏亞硫酸鉀水溶液消毒後，再用熱水裡外沖洗掉偏亞硫酸鉀水。
13. 利用虹吸管，將新釀啤酒裝滿啤酒瓶，不要裝的太滿，餘留大約2吋的空間。加蓋密封。
14. 放置在8°C左右陰暗的地方，熟成2週。

Secondary fermentation and clarification

9. Boil two cups of water with 3/4 cup of fructose in pan.
10. Dissolve gelatin in 300ml. of cold water, heat in a double-boiler until transparent.
11. Pour the sugar solution and gelatin into the finished beer, stir with a stirring paddle until evenly mixed.
12. Rinse the empty beer bottles thoroughly to get rid of the potassium metabisulfite solution after removing from the potassium metabisulfite solution for sterilizing.
13. Siphon with a siphon tube and fill the empty beer bottles with finished beer. Do not fill up all the way, leave about 2 inches of space, then seal with a cap.
14. Let age in a cool dark place at about 8C for 2 weeks.

German Lager
德國式啤酒

- 製成份量：約0.6公升×30瓶
- 需要時間：8週

基本風格的德國式啤酒，顏色有一點棕紅色稍甜，淡淡的啤酒花苦味。

A basic German beer. The color is a little reddish brown, the flavor slightly sweet with a light bitter overtone provided by the beer hops.

材料：

Premium Lager	
濃縮麥芽汁	1罐（2公斤）
Amber dry濃縮麥芽粉	1公斤
Hallertau啤酒花顆粒	15公克（苦味）
Tettnang啤酒花顆粒	15公克（芳香）
Lager啤酒酵母菌	1包
果糖	3/4杯（180ml.）
素食用吉利丁	5公克

- 第一次加入的啤酒花是要萃取啤酒花的苦味，關火前或關火後再加入的啤酒花，則為了萃取啤酒花的香味。

Ingredients

1 can (2kg) Premium Lager malt extract liquid
1kg Amber dry malt extract
15g Hallertau hops pellets (bitter agent)
15g Tettnang hops pellets (aroma)
1 pack Lager yeast
3/4C (180ml.) fructose
5g vegetarian gelatin

- Portion Preparation: approximately 0.6 liters x 30 bottles
- Time Needed: 8 weeks

* The first hops is for its bitterness and the second hops added just before or after removing from the heat is for its aroma.

麥芽汁

麥芽粉

啤酒花顆粒

啤酒酵母菌

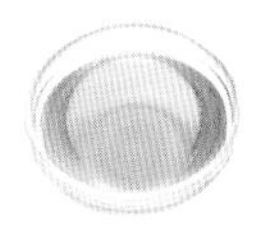
果糖

吉利丁

German Lager

前製作業

取一洗淨的塑膠桶，放入4公升的水及60公克的偏亞硫酸鉀，充份溶解後把要用的工具，如長柄匙、發酵鎖、虹吸管、濾網、比重測糖計等，全部泡在裡面。

做法

1. 不銹鋼大鍋中倒入6公升的水煮開，加入Premium Lager濃縮麥芽汁和Hallertau啤酒花顆粒一起煮沸20分鐘，再加入Tettnany啤酒花繼續煮5分鐘熄火，製成麥芽汁。
2. 將10公升冰開水倒入玻璃發酵瓶中，將麥芽汁過濾倒入。
3. 添加冷水，直到瓶中有19公升的麥芽汁。加入Amber dry麥芽濃縮粉，攪拌溶解。
4. 取出半杯麥芽汁，放冷後（約16°C）用比重測糖計量取糖的含量，紀錄在筆記本。
5. 當麥芽汁降溫到24°C時，加入Lager酵母菌，以長柄匙攪拌均勻。
6. 放置在20°C左右的陰暗處發酵。
7. 24小時後檢查，發酵鎖有氣泡不斷的湧出表示發酵正常。將發酵瓶移到8°C左右陰暗處，繼續發酵。
8. 低溫發酵4週後，當發酵作用完成，發酵鎖不再有氣泡冒出（也許1分鐘才冒出一個氣泡，這種狀況也算發酵完成），倒出一些麥芽汁以比重測糖計檢測含糖量，降到0度左右時即表示第一次發酵作用完成，虹吸換瓶。

Preparation

In a clean plastic bucket combine 4 liters of water with 60 grams of potassium metabisulfite, stir until dissolved completely, then soak all the necessary utensils, such as long handled stirring paddle, airlocks, siphon tubes, sieve and hydrometer, in the solution.

Methods:

1. To prepare wort: Boil 6 liters of water in a stainless steel pan, add Premium Lager malt extract liquid and Hallertau hops pellets, then bring to a rolling boil for 20 minutes. Add Tettnany hops pellets and continue boiling for 5 minutes and remove from heat.
2. Fill 25 liter glass fermeater with 10 liters of ice-cold water. Pour wort into it through a sieve to remove any dregs.
3. Add cold water as needed until the fermenter holds 19 liters of liquid. Add Amber dry malt extract, stir until well-dissolved.
4. Remove half cup of wort and let cool (around 16C). Measure with a hydrometer to test the sugar gravity and record.

5. When the temperature of the wort is down to 24C, pitch in Lager yeast and stir with a stirring paddle until well-mixed.
6. Let ferment in a dark place at around 20C.
7. Wait for 24 hours and check if the fermentation has started. If bubbles keep foaming out of the airlocks, it shows that the fermentation is going on normally.
8. Let ferment for about 4 weeks. When the bubbles finally stop foaming out the airlocks (sometimes it takes a minute for the foams to bubble out, but that still counts as finished), pour out a little wort and measure with a hydrometer to test the sugar gravity. If it is around 0, it means that the primary fermentation has finished. Siphon and bottle.

第二次發酵及澄清處理

9. 將兩杯水與3/4杯果糖一起放入鍋中煮開。
10. 吉利丁溶解於300 ml.冷水中，隔水加熱成透明狀。
11. 糖水和吉利丁液倒入新釀啤酒中，以長柄匙攪拌均勻。
12. 將啤酒瓶浸泡在偏亞硫酸鉀水溶液消毒後，再用熱水裡外沖洗掉偏亞硫酸鉀水。
13. 利用虹吸管，將新釀啤酒裝滿啤酒瓶，不要裝的太滿，餘留大約2吋的空間。加蓋密封。
14. 放置在8℃左右陰暗的地方，熟成3週。

Secondary fermentation and clarification

9. Boil two cups of water with 3/4 cup of fructose in pan.
10. Dissolve gelatin in 300ml. of cold water, heat in a double-boiler until transparent.
11. Pour the sugar solution and gelatin in the finished beer, stir with a stirring paddle until evenly mixed.
12. Rinse the empty beer bottles thoroughly to get rid of the potassium metabisulfite solution after removed from the potassium metabisulfite solution for sterilizing.
13. Siphon with a siphon tube and fill the empty beer bottles with finished beer. Do not fill up all the way, leave about 2 inches of space, then seal with a cap.
14. Let age at cool dark place about 8C for 3 weeks.

Irish Stout
愛爾蘭苦啤酒

- 製成份量：約0.6公升×30瓶
- 需要時間：6週

可以說是愛爾蘭的國飲，又黑又苦，濃郁複雜的風味頗有愛爾蘭的特色。

The anthem of Ireland, dark and bitter, whose thick flavor is reminiscent of Ireland.

材料：

Irish stout濃縮麥芽汁	1罐（2公斤）
Dark dry濃縮麥芽粉	1公斤
Bullion Leaf啤酒花顆粒	25公克（苦味）
Stout啤酒酵母菌	1包
果糖	3/4杯（180ml.）
素食用吉利丁	5公克

- Bullion Leaf是苦澀味特別濃的啤酒花顆粒。

Ingredients

1 can (2kg) Irish Stout malt extract liquid
1kg Dark dry malt extract
25g Bullion Leaf hops pellets
1 pack Stout yeast
3/4C (180ml.) fructose
5g vegetarian gelatin

- Portion Preparation: approximately 0.6 liters x 30 bottles
- Time Needed: 6 weeks

* Bullion Leaf is hops pellets with the strongest bitterness.

麥芽汁

麥芽粉

啤酒花顆粒

啤酒酵母菌

果糖

吉利丁

Irish Stout

前製作業

取一洗淨的塑膠桶，放入4公升的水及60公克的偏亞硫酸鉀，充份溶解後把要用的工具，如長柄匙、發酵鎖、虹吸管、濾網、比重測糖計等，全部泡在裡面。

做法

1. 不銹鋼大鍋中倒入6公升的水煮開，加入Irish stout濃縮麥芽汁和Bullion Leaf煮沸30分鐘，製成麥芽汁。
2. 將10公升冰開水倒入玻璃發酵瓶中，將麥芽汁過濾倒入。
3. 添加冷水，直到瓶中有19公升的麥芽汁。加入Dark dry濃縮麥芽粉，攪拌溶解。
4. 取出半杯麥芽汁，放冷後（約16℃）用比重測糖計量取糖的含量，紀錄在筆記本。
5. 當麥芽汁降溫到24℃時，加入Stout酵母菌，以長柄匙攪拌均勻。
6. 放置在20℃左右的陰暗處發酵。
7. 24小時後檢查發酵作用是否開始，發酵鎖有氣泡不斷的湧出表示發酵正常。
8. 約15天，當發酵鎖湧出的氣泡逐漸減弱而停止時，再倒出一些麥芽汁以比重測糖計檢測含糖量，降到0度左右時即表示第一次發酵作用完成，虹吸換瓶。

Preparation

In a clean plastic bucket combine 4 liters of water with 60 grams of potassium metabisulfite, stir until dissolved completely then soak all the necessary utensils, such as long handled stirring paddle, airlocks, siphon tubes, sieve and hydrometer, in the solution.

Methods:

1. To prepare wort: Bring 6 liters of water to a boil, add Irish Stout malt extract liquid and Bullion Leaf, then boil vigorously for 30 minutes.
2. Fill 25 liter glass fermeater with 10 liters of ice-cold water. Pour wort into it through a sieve to remove any dregs.
3. Add cold water as needed until the fermenter holds 19 liters of liquid. Add Dark dry malt extract, stir until well-dissolved.
4. Remove half cup of wort and let cool (around 16C). Measure with a hydrometer to test the sugar gravity and record.
5. When the temperature of the wort is down to 24C, pitch in Stout yeast and stir with a stirring paddle until well-mixed.

6 . Let ferment in a dark place at around 20C.
7 . Wait for 24 hours and check if the fermentation has started. If bubbles keep foaming out of the airlocks, it shows that the fermentation is going on normally.
8 . Let ferment for about 15 days. When the bubbles finally stop foaming out the airlocks, pour out a little wort and measure with a hydrometer to test the sugar gravity. If it is around 0, it means that the primary fermentation has finished, siphon and bottle.

第二次發酵及澄清處理

9 . 將兩杯水與3/4杯果糖一起放入鍋中煮開。
10 . 吉利丁溶解於300 ml.冷水中，隔水加熱成透明狀。
11 . 糖水和吉利丁液倒入新釀啤酒中，以長柄匙攪拌均勻。
12. 將啤酒瓶浸泡在偏亞硫酸鉀水溶液消毒後，再用熱水裡外沖洗掉偏亞硫酸鉀水。
13 . 利用虹吸管，將新釀啤酒裝滿啤酒瓶，不要裝的太滿，餘留大約2吋的空間。加蓋密封。
14 . 放置在8°C左右陰暗的地方，熟成4週。

Secondary fermentation and clarification

9. Boil two cups of water with 3/4 cup of fructose in pan.
10. Dissolve gelatin in 300ml. of cold water, heat in a double-boiler until transparent.
11. Pour the sugar solution and gelatin in the finished beer, stir with a stirring paddle until evenly mixed.
12. Rinse the empty beer bottles thoroughly to get rid of the potassium metabisulfite solution after removed from the potassium metabisulfite solution for sterilizing.
13. Siphon with a siphon tube and fill the empty beer bottles with finished beer. Do not fill up all the way, leave about 2 inches of space, then seal with a cap.
14. Let age in a cool dark place at about 8C for 4 weeks.

California Style Low Calorie Light Beer
加州式低熱量淡啤酒

■ 製成份量：約0.6公升×30瓶
■ 需要時間：6週

低卡洛里減肥食品也延伸到日常飲用的啤酒，流行於美國的這款低熱量淡啤酒，釀製時減少麥芽的使用量，改用稻米的濃縮汁，口味清淡，酒精含量低，是炎炎夏日想要喝啤酒又怕體重增加的應時產品。

The low-calorie product trend has finally overtaken every-day beers. This is a popular low-calorie beer in America. The amount of barley used in brewing has been reduced in favor of concentrated rice. The flavor is light with a low alcohol content. It is a perfect drink for hot summer days when you are sensitive to possible weight gain.

材料：

Pale Malt濃縮麥芽汁	半罐（1公斤）
稻米濃縮汁	1.5公斤
Lager啤酒酵母菌	1包
果糖	3/4杯（180ml.）
素食用吉利丁	5公克

■ 加州品牌Pale Malt濃縮麥芽汁已經加入了啤酒花，所以不必再添加。

Ingredients

1 can (2kg) Pale Malt extract liquid
1.5kg rice liquid extract
1 pack Lager yeast
3/4C (180ml.) fructose
5g vegetarian gelatin
* California brand Pale Malt extract liquid already has hops added, therefore hops are not needed in this recipe.

■ Portion Preparation: approximately 0.6 liters x 30 bottles
■ Time Needed: 6 weeks

麥芽汁

稻米濃縮汁

啤酒酵母菌

果糖

吉利丁

California Style Low Calorie Light Beer

前製作業

取一洗淨的塑膠桶，放入4公升的水及60公克的偏亞硫酸鉀，充份溶解後把要用的工具，如長柄匙、發酵鎖、虹吸管、濾網、比重測糖計等，全部泡在裡面。

做法

1. 不銹鋼大鍋中倒入6公升的水煮開，加入Pale Malt濃縮麥芽汁繼續煮20分鐘，放入稻米濃縮汁後熄火，製成麥芽汁。
2. 將10公升冰開水倒入玻璃發酵瓶中，將麥芽汁過濾倒入。
3. 添加冷水，直到瓶中有19公升的麥芽汁。
4. 取出半杯麥芽汁，放冷後（約16℃）用比重測糖計量取糖的含量，紀錄在筆記本。
5. 當麥芽汁降溫到24℃時，加入Lager啤酒酵母，以長柄匙攪拌均勻。
6. 第一天發酵時，溫度保持在22℃左右。
7. 24小時後將溫度降至10℃左右，繼續發酵。
8. 約14天，當發酵鎖湧出的氣泡逐漸減弱而停止時，再倒出一些麥芽汁以比重測糖計檢測含糖量，降到0度左右時即表示發酵作用完成。
9. 將啤酒瓶浸泡在偏亞硫酸鉀水溶液消毒後，再用熱水裡外沖洗掉偏亞硫酸鉀水。
10. 利用虹吸管，將新釀啤酒裝滿啤酒瓶，不要裝的太滿，餘留大約2吋的空間。加蓋密封。

※ 請注意：這款啤酒並不需要第二次發酵處理。

Preparation

In a clean plastic bucket combine 4 liters of water with 60 grams of potassium metabisulfite, stir until dissolved completely, then soak all the necessary utensils, such as the long handled stirring paddle, airlocks, siphon tubes, sieve and hydrometer, in the solution.

Methods:

1 . To prepare wort: Boil 6 liters of water in a stainless steel pan, add Pale Male extract liquid and continue boiling for 20 minutes. Add rice liquid extract and remove from heat.

2. Fill 25 liter glass fermeater with 10 liters of ice-cold water. Pour wort into it through a sieve to remove any dregs.

3 . Add cold water as needed until the fermenter holds 19 liters of liquid. Add plain dry malt extract, stir until well-dissolved.

4 . Remove half cup of wort and let cool (around 16C). Measure with a hydrometer to test the sugar gravity and record.

5 . When the temperature of the wort is down to 24C, pitch in Lager yeast and stir with a stirring paddle until well-mixed.

6 . Let ferment in a dark place at around 22C.

7 . Wait for 24 hours, then lower the temperature to 10C.

8 . Let ferment for about 14 days. When the bubbles finally stop foaming out the airlocks, pour out a little wort and measure with a hydrometer to test the sugar gravity. If it is around 0, it means that the primary fermentation has finished

9 . Rinse the empty beer bottles thoroughly to get rid of the potassium metabisulfite solution after removed from the potassium metabisulfite solution for sterilizing.

10 . Siphon with a siphon tube and fill the empty beer bottles with finished beer. Do not fill up all the way, leave about 2 inches of space, then seal with a cap.

* Note that: This beer does not require secondary fermentation.

Pilsner

捷克式
重口味啤酒

■ 製成份量：約0.6公升×30瓶
■ 需要時間：10週

捷克啤酒和德國啤酒的釀製方法相同，採用底部發酵的酵母菌。但捷克式風格啤酒的顏色是呈淡琥珀色，有著較濃的麥芽香，充滿了辛辣的啤酒花風味。

The brewing method is exactly the same as the German beer, which uses bottom-fermenting yeast. However, Pilsner is a light-amber color with a thick barley aroma and a spicy hops flavor.

材料：

捷克Pilsner濃縮麥芽汁	1罐（2公斤）
Dutch extra light dry malt 濃縮麥芽粉	1公斤
Saaz啤酒花顆粒	15公克（苦味）
Saaz啤酒花顆粒	10公克（芳香）
Lager啤酒酵母菌	1包
果糖	3/4杯（180ml.）
素食用吉利丁	5公克

■ 第一次加入的啤酒花是要萃取啤酒花的苦味，關火前或關火後再加入的啤酒花，則為了萃取啤酒花的香味。

Ingredients

1 can (2kg) Pilsner Ale malt extract liquid
1kg Dutch Extra light dry walt
15g saaz hops pellets
1 pack Lager yeast
3/4C (180ml.) fructose
5g vegetarian gelatin

■ Portion Preparation: approximately 0.6 liters x 30 bottles
■ Time Needed: 4 weeks

麥芽汁

麥芽粉

啤酒花顆粒

啤酒酵母菌

果糖

吉利丁

Pilsner

前製作業

取一洗淨的塑膠桶，放入4公升的水及60公克的偏亞硫酸鉀，充份溶解後把要用的工具，如長柄匙、發酵鎖、虹吸管、濾網、比重測糖計等，全部泡在裡面。

做法

1. 不銹鋼大鍋中倒入6公升的水煮開，加入Pilsner濃縮麥芽汁煮沸20分鐘，放入15公克Saaz啤酒花顆粒續煮10分鐘，再加入10公克Saaz啤酒花煮1分鐘熄火，製成麥芽汁。
2. 將10公升冰開水倒入玻璃發酵瓶中，將麥芽汁過濾倒入。
3. 添加冷水，直到瓶中有19公升的麥芽汁。加入Dutch extra light dry malt濃縮麥芽粉，攪拌溶解。
4. 取出半杯麥芽汁，放冷後（約16°C）用比重測糖計量取糖的含量，紀錄在筆記本。
5. 當麥芽汁降溫到24°C時，加入Lager啤酒酵母菌，以長柄匙攪拌均勻。
6. 放置在6°C左右的陰暗處緩慢發酵。
7. 24小時後檢查發酵作用是否開始，發酵鎖有氣泡不斷的湧出表示發酵正常。
8. 約4～5週後，當發酵鎖湧出的氣泡逐漸減弱而停止時，再倒出一些麥芽汁以比重測糖計檢測含糖量，降到0度左右時即表示第一次發酵作用完成，虹吸換瓶。

Preparation

In a clean plastic bucket combine 4 liters of water with 60 grams of potassium metabisulfite, stir until dissolved completely then soak all the necessary utensils, such as the long handled stirring paddle, airlocks, siphon tubes, sieve and hydrometer, in the solution.

Methods:

1. To prepare wort: Boil 6 liters of water in a stainless steel pan, add Pilsner malt extract liquid. Boil vigorous for 15 minutes then add saaz hops pellets before removing from heat.
2. Fill 25 liter glass fermeater with 10 liters of ice-cold water. Pour wort into it through a sieve to remove any dregs.
3. Add cold water as needed until the fermenter holds 19 liters of liquid. Add Dutch extra light dry malt extract, stir until well dissolved.
4. Remove half cup of wort and let cool (around 16C). Measure with a hydrometer to test the sugar gravity and record.

5 . When the temperature of the wort is down to 24C, pitch in the yeast and stir with a stirring paddle until well-mixed.
6 . Let ferment in a dark place at around 20C.
7 . Wait for 24 hours and check if the fermentation has started. If bubbles keep foaming out of the airlocks,, it shows that the fermentation is going on normally.
8 . Let ferment for about 1 week. When the bubbles finally stop foaming out the airlocks, pour out a little wort and measure with a hydrometer to test the sugar gravity. If it is around 0, it means that the primary fermentation has finished, siphon and bottle.

第二次發酵及澄清處理

9.將兩杯水與3/4杯果糖一起放入鍋中煮開。
10.吉利丁溶解於300 ml.冷水中，隔水加熱成透明狀。
11.糖水和吉利丁液倒入新釀啤酒中，以長柄匙攪拌均勻。
12.將啤酒瓶浸泡在偏亞硫酸鉀水溶液消毒後，再用熱水裡外沖洗掉偏亞硫酸鉀水。
13.利用虹吸管，將新釀啤酒裝滿啤酒瓶，不要裝的太滿，餘留大約2吋的空間。加蓋密封。
14.放置在8°C左右陰暗的地方，熟成4週。

Secondary fermentation and clarification

9 . Boil two cups of water with 3/4 cup of fructose in pan.
10 . Dissolve gelatin in 300ml. of cold water, heat in a double-boiler until transparent.
11 . Pour the sugar solution and gelatin in the finished beer, stir with a stirring paddle until evenly mixed.
12 . Rinse the empty beer bottles thoroughly to get rid of the potassium metabisulfite solution after removing from the potassium metabisulfite solution for sterilizing.
13 . Siphon with a siphon tube and fill the empty beer bottles with finished beer. Do not fill up all the way, leave about 2 inches of space, then seal with a cap.
14 . Let age in a cool dark place at about 8C for 4 weeks.

Mexican Style Beer

墨西哥式啤酒

■ 製成份量：約0.6公升×30瓶
■ 需要時間：8～10週

微甜的麥芽香，淡淡的啤酒花苦味，聞名世界的墨西哥啤酒要歸功於百年前自奧地利移民至此的釀酒師帶來的歐洲釀酒技術。

With slightly barley aroma and light bitter hop flavor this famous Mexican beer has its roots in the brewing culture brought to Mexico by Austrian immigrants.

材料：

Dutch dark Larger麥芽汁 1罐（2公斤）
已摻入啤酒花的麥芽汁 半罐（1公斤）

Tettnang啤酒花顆粒 145公克
Lager啤酒酵母菌 1包
果糖 3/4杯（180ml.）

■ 加州品牌濃縮麥芽汁已經加入了啤酒花，所以不必再添加。

Ingredients

1 can (2kg) Dutch dark Larger malt extract liquid
1/2 can light malt extract
15g Tettnang hops pellets
1 pack Lager yeast
3/4C (180ml.) fructose
5g vegetarian gelatin

■ Portion Preparation: approximately 0.6 liters x 30 bottles
■ Time Needed: 8~10 weeks

麥芽汁

麥芽粉

啤酒花顆粒

啤酒酵母菌

果糖

Mexican

前製作業

取一洗淨的塑膠桶，放入4公升的水及60公克的偏亞硫酸鉀，充份溶解後把要用的工具，如長柄匙、發酵鎖、虹吸管、濾網、比重測糖計等，全部泡在裡面。

做法

1. 不銹鋼大鍋中倒入6公升的水煮開，加入Dutch dark Larger麥芽汁和已摻入啤酒花的麥芽汁繼續煮20分鐘，放入Tettnang啤酒花增加香氣，隨即熄火，燜5分鐘，製成麥芽汁。
2. 將10公升冰開水倒入玻璃發酵瓶中，將麥芽汁過濾倒入。
3. 添加冷水，直到瓶中有19公升的麥芽汁。
4. 取出半杯麥芽汁，放冷後（約16°C）用比重測糖計量取糖的含量，紀錄在筆記本。
5. 當麥芽汁降溫到24°C時，加入Lager啤酒酵母，以長柄匙攪拌均勻。
6. 放置在8°C左右的陰暗處緩慢發酵。
7. 4小時後檢查發酵作用是否開始，發酵鎖有氣泡不斷的湧出表示發酵正常。
8. 約4～6週後，當發酵鎖湧出的氣泡逐漸減弱而停止時，再倒出一些麥芽汁以比重測糖計檢測含糖量，降到0度左右時即表示第一次發酵作用完成，虹吸換瓶。

Preparation

In a clean plastic bucket combine 4 liters of water with 60 grams of potassium metabisulfite, stir until dissolved completely then soak all the necessary utensils, such as the long handled stirring paddle, airlocks, siphon tubes, sieve and hydrometer, in the solution.

Methods:

1. To prepare wort: Boil 6 liters of water in a stainless steel pan, add Dutch dark Larger malt extract liquid and light malt extract. Boil vigorously for 15 minutes, then add Tettnang hops pellets before removing from heat.
2. Fill 25 liter glass fermeater with 10 liters of ice-cold water. Pour wort into it through a sieve to remove any dregs.
3. Add cold water as needed until the fermenter holds 19 liters of liquid. Add plain dry malt extract, stir until well-dissolved.
4. Remove a half cup of wort and let cool (around 16C). Measure with a hydrometer to test the sugar gravity and record.
5. When the temperature of the wort is down to 24C, pitch in the yeast and stir with a stirring paddle until well-mixed.

6 . Let ferment in a dark place at around 20C.
7 . Wait for 24 hours and check if the fermentation has started. If bubbles keep foaming out of the airlocks, it shows that the fermentation is going on normally.
8 . Let ferment for about 1 week. When the bubbles finally stop foaming out the airlocks, pour out a little wort and measure with a hydrometer to test the sugar gravity. If it is around 0, it means that the primary fermentation has finished, siphon and bottle.

Brewing

第二次發酵及澄清處理

9. 將兩杯水與3/4杯果糖一起放入鍋中煮開。
10. 吉利丁溶解於300 ml.冷水中，隔水加熱成透明狀。
11. 糖水和吉利丁液倒入新釀啤酒中，以長柄匙攪拌均勻。
12. 將啤酒瓶浸泡在偏亞硫酸鉀水溶液消毒後，再用熱水裡外沖洗掉偏亞硫酸鉀水。
13. 利用虹吸管，將新釀啤酒裝滿啤酒瓶，不要裝的太滿，餘留大約2吋的空間。加蓋密封。
14. 放置在8°C左右陰暗的地方，熟成4週。

Secondary fermentation and clarification

9 . Boil two cups of water with 3/4 cup of fructose in pan.
10 . Dissolve gelatin in 300ml. of cold water, heat in a double-boiler until transparent.
11 . Pour the sugar solution and gelatin in the finished beer, stir with a stirring paddle until evenly mixed.
12 . Rinse the empty beer bottles thoroughly to get rid of the potassium metabisulfite solution after removing from the potassium metabisulfite solution for sterilizing.
13 . Siphon with a siphon tube and fill the empty beer bottles with finished beer. Do not fill up all the way, leave about 2 inches of space, then seal with a cap.
14 . Let age in a cool dark place at about 8C for 4 weeks.

Brewing

啤酒色香味的品評

釀製啤酒的原料，不外乎麥芽及少量的小麥、玉米、稻米等添加物，再加上啤酒花，啤酒酵母及水，是世界上最古老的酒精飲料，考古學家發現在9仟年前，歐洲居民就已經知道用大麥浸水發芽，自然發酵製成麥酒，到了古埃及文明更將此種釀酒技術加以改進，將大麥磨粉烘烤成調包加水混合發酵，再加入植物的根莖葉及蜂蜜調味，研發出十數種不同口味的啤酒。

喝啤酒講求的是將一大杯冰涼的啤酒大口大口的喝下肚，似乎很少有人像喝葡萄酒似的玩賞。品評它的色香味，其實啤酒的好壞有很大的差別，必需具備有色香味的基本條件及正確的儲存方法，我們才能享受到美味的啤酒。

儲存的時間對啤酒有決定性的影響，除了少數特釀啤酒需要長時間的熟成，大多數的啤酒都要趁新鮮時喝，放置時瓶蓋向上，直立在紙箱中，放在低溫黑暗沒有光線的地方，大家是否注意到啤酒瓶都是深色的玻璃瓶？就是為了保存啤酒的新鮮度減少光線的照射。

剛從冰箱拿出來的啤酒如果溫度太低，最好先倒入啤酒杯中再喝，湧出的氣泡將啤酒的香味充份揮發，否則溫度過低的啤酒喝不出來它的芳香，淡啤酒最適合的溫度是13°C左右，色濃味厚的啤酒16°C時更加的香醇有味，喝啤酒時使用的玻璃杯一定要乾淨沒有油漬、髒污或是殘餘的洗潔精，否則啤酒倒入不清潔的杯裡，會減少泡沫的形成，一杯清涼的啤酒沒有滿溢的泡沫，不但損失了視覺的觀賞更缺乏了躍動的清涼感，多麼的掃興！

釀製啤酒時，使用的麥芽因為烘焙時間的不同而影響到啤酒的色澤。自淡黃的稻草色、金黃色、琥珀色到深濃的黑褐色，品評啤酒的顏色最主要就是觀看它的顏色是否符合它的品種特色。換句話說，美國式淡啤酒卻有琥珀的色澤就是一大敗筆。

手持啤酒杯，輕微傾斜約30° ，將啤酒緩緩倒入杯中。啤酒表面的泡沫將啤酒中的香味揮發出來，此時深深的吸一口氣，聞到的是啤酒花香？麥芽香？還是酒香？應該是各種香的混合體，Ale啤酒酵母在發酵時會產生水果的清香，Lagar啤酒酵母則會賦予啤酒辛香的味道，不同品種及產地的啤酒花也有各種不同的香味，最常表現在啤酒裡的是花香、果香，或是森林的樹木香味：香和味是一體兩面鼻子聞到的是香，口中品嚐的是味，如果感冒時鼻塞，這時口中吃到的食物均沒有了味道，喝啤酒時先別急著嚥下去，讓啤酒在口中停留一下，接觸到舌頭前後左右的味蕾，品味一下是否有麥芽的甜味？啤酒花的香及苦澀。下一次，當你喝啤酒時，別忘了細細品味一下啤酒的色、香、味。

Beer tasting: the aroma, the color and the flavor

The ingredients of beer brewing are generally barley, wheat, corn or grains, along with hops, yeast and water. It is the oldest alcoholic drink known to mankind. Archaeologists have discovered that nine thousand years ago, Europeans already knew how to soak barley in water until it sprouted, then let it ferment naturally to brew beer. Ancient Egypt improved on this brewing technique by using roasted ground barley combined with water. The combined ingredients were fermented first, then plant roots or leaves and honey are added to season. In this fashion many flavors of beer were invented.

When serving beer, the correct fashion is to drink the icy cold beer in one big gulp after another. Nobody serves beer the way wine is served. Despite the variation in taste, color, and aroma, there is a major difference between good beer and bad beer. If the beer has the right qualifications, such as color, aroma and taste, and is stored properly, we can really enjoy a bottle of good beer.

The duration of storage of the beer is a decisive influence on the beer. Except for some beers which have to be aged in a dark place for a long time, most beer has to be served while still fresh. Store the beer upright with the cap facing up in a cool dark place. Have you noticed that the beer bottles are all dark glass? This helps preserve the freshness of the beer by reducing exposure to sunlight.

When the beer just out of the refrigerator is too cold, serve it in a beer mug. The beer aroma will fully evaporate by the poping foams. Otherwise, the beer which is too cold is difficult to taste its aroma. The most suitable temperature for the light beer is around 13C and thick dark beer is even better around 16C. Use a clean glass without any oily stains, dirt or remaining dish detergent when serving beer. Or when the beer is poured in a unclean glass, the foams will be so forth decreased. A glass of icy cold beer without foams poping around the edge, it is really a shame for the eye sight to lose its enjoyment, it is also lacking of leaping rhyme for the feeling.

When brewing beer, the roasting time of the barley influences the color of the beer. Whether a light straw color, golden, amber, or a thick dark brown color, the most important thing is to taste the beer to see if the color of the beer suits its type and characteristics. In another words, American light beer with an amber color is a failure.

Hold the beer mug slightly tilted, at about a 30 degree angle, and pour the beer gently into the mug. The foam on the surface of the beer should release the aroma of the beer. Now breathe deep and inhale the aroma. Is it a flowery aroma? A barley aroma? or an alcoholic aroma? It should be the combination of all. Ale produces a fruity aroma when the yeast is fermenting. Lager produces an alcohol aroma. Different brands and origins of the hops also yield different aromas. The most common aromas are the flowery aroma, fruity aroma or an aroma of trees like a forest. Aroma and taste are the two sides of the same coin. When the nose detects an aroma, the mouth tastes a flavor. Hence if you catch a cold, everything to you is tasteless. When serving beer, do not try to swallow it in a hurry. Let the beer stay in your mouth for a little while, feeling the taste on your tongue from all sides. Taste it carefully to see if it is sweet like the barley, or fragrant and bitter like the hops. Next time, when you serve your beer, do not forget use your newfound awareness to taste the color, aroma and flavor of the beer.

Cooking with Beers

Chapter III

香草雞胸肉排Chicken Steak with Herbs
麻油雞Sesame Chicken with Beer
燜燒牛肉Simmered Beef
啤酒醃牛排Beef Steak with Beer
蒜烤羊排Garlic Roasted Lamb Ribs
啤酒芥末火腿Beef Steak with Beer
烤甜杏夾心豬排Apricot-Filled Pork Rib Chop
啤酒白汁肉丸White Sauce Meatballs with Beer
紅燒豬腳Red-Cooked Pork Feet
煎鱈魚排Fried Cod Steak
清蒸草蝦Steamed Grass Shrimp
啤酒蝦Beer Shrimp
啤酒松子醬Beer Pine Nut Dressing
涼拌白菜心Cold Mixed Napa Cabbage with Vinegar and Beer
炒青菜Stir-Fried Kale
舊金山海鮮湯San Francisco Seafood Soup

3 應用篇

用啤酒做中西佳餚

PART THREE

應用篇-用啤酒做中西佳餚

1970年初，我搬到美國加州定居，用米酒做菜的習慣，不得不入境隨俗，跟著老美用雪莉酒（Sherry）當料理酒；醃肉、燒魚都差強人意，但是做成麻油雞就不對味了。30年前我們居住的小鎮根本買不到東方式的米酒，窮則變、變則通，我用台灣帶來的黑麻油炒香了薑片和雞塊，加入整罐的啤酒，結果熬出來的麻油雞竟然滑嫩好吃的不得了！連挑嘴的老公和婆婆都讚不絕口。

從此「啤酒」成了我料理檯上的秘密武器，烹調出一道道吃了還想再吃的獨門口味，所以在這本自製啤酒的書中，我也將我的啤酒料理現出，和讀者一起分享！

會將啤酒運用到西餐料理完全是個偶然，家中的3個小孩到了念初中的年齡，時常要求我做一些西式晚餐，可是先生卻偏愛中式料理，所以我將傳統西式料理稍加變化，做成了有點西又有點中的改良菜，且大量加入啤酒入菜，烹調出一家人都能接受的西式佳餚。

In the early 1970s, I settled down in California. The rice wine I used to cook with had to be substituted with the sherry that most Americans use as a cooking wine. It was quite satisfactory when it came to marinating pork or cooking with fish. However, when it came to cooking the sesame chicken soup, it was just not the flavor I expected.

Thirty years ago, in the small town where we lived, no place carried eastern rice wines. When it came to cook, I used the black sesame oil that I brought with me from Taiwan to stir-fry the ginger slices and chicken pieces until the flavor was released, then I poured in a whole can of beer. The result was that the chicken turned out to be quite smooth and tender. Even my picky husband and mother-in-law could not stop praising the dish.

From that time on, beer became my secret weapon at my kitch. With beer, delicious, unique dishes came out one after another, so good that you could not help but ask for more. Therefore, I have written this recipes to share my beer cuisine with my readers!

My use of beer in western cuisine is purely an accident. When my 3 kids entered middle school, they would once a while would ask me to prepare a western meal for dinner. However, my husband usually wanted me to cook something Chinese. Therefore, I changed the traditional western cuisine slightly to make it a hybrid of Chinese and western style. Also with a large amount of beer added, the meal became food that was acceptable to all parties.

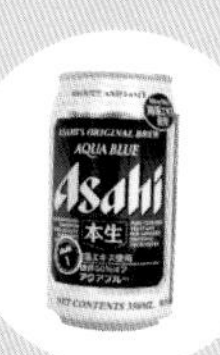

Chicken Steak with Herbs
香草雞胸肉排

做法

1. 雞胸肉去皮去骨槌成薄片。
2. 平底鍋加入1大匙橄欖油及1大匙奶油炒香大蒜、月桂葉及百里香。
3. 加入雞胸肉兩面煎黃，淋入啤酒和白蘭地。加鹽和胡椒調味，待湯汁收乾後即可起鍋。
4. 鍋內加1/2杯水，以鍋底的焦香煮成淋汁，加入1/2匙太白粉調勻，淋在雞排上即成。

■ 去皮的雞胸肉沒有什麼油脂，且肉質細嫩，藉著香草和酒的提味，使得平淡的雞胸肉好吃得不得了。

材料：

雞胸肉	4片
蒜碎	3大匙
月桂葉	2片
百里香末	1/2小匙
啤酒	60 ml.
白蘭地	2大匙

調味料：

鹽	1大匙
胡椒	1/2小匙

Methods:

1. Remove and discard the skin from chicken breast and tenderize until thin with a tenderizer.
2. Heat 1T. of olive oil and 1T. of butter in frying pan, stir-fry minced garlic, bay leaves and thyme until fragrant.
3. Add chicken breasts and fry until golden on both sides. Drizzle with beer and brandy. Season with salt and pepper to taste. Cook until the liquid is absorbed and remove from heat.
4. Add 1/2C of water to the frying pan and cook until the flavor on the bottom has combined with the water for drizzling. Thicken with 1/2 spoonful of cornstarch and drizzle over the chicken steaks. Serve.

■ Skinless chicken breast has not much fat, however the meat is tender and juicy fine. With the enhancement of herbs and beer, this ordinary chicken breast suddenly becomes quite enjoyable.

Ingredients:

4 chicken breasts, 3T. minced garlic, 2 bay leaves, 1/2t. minced thyme, 60ml. beer, 2T. brandy

Marinade:

1T. salt, 1/2t. pepper

Sesame Chicken with Beer
麻油雞

做法

1. 雞洗淨切塊。
2. 起油鍋放入黑麻油，爆香薑片，加入雞塊翻炒，至雞塊呈淡黃色。
3. 倒入啤酒，煮開後轉小火繼續燉煮約30分鐘即可上桌。

■ 啤酒經過燉煮，原有的酒精成份都揮發了，麻油雞湯不僅味道鮮美且帶著麥芽的香甜味，非常好吃。

材料：

土雞	1/2隻（約1斤半）
老薑	6大片
黑麻油	4大匙
啤酒	1瓶（600ml.）

Methods:

1. Rinse chicken well and cut into pieces.
2. Heat wok first, add black sesame oil until smoking. Stir-fry ginger slices until fragrant, add chicken pieces to mix. Saute rapidly until the chicken pieces are light brown.
3. Pour in a whole bottle of beer and bring to a boil, then reduce heat to low. Continue cooking for about 30 minutes until done. Remove and serve.

■ The alcohol in beer evaporates after extended cooking, leaving the sesame chicken soup fresh and delicious with a sweet aroma. This dish is wonderful.

Ingredients:

1/2 free range chicken (approximately 2lbs.), 6 large ginger slices, 4T. black sesame oil, 1 bottle beer (600ml.)

Simmered Beef
燜燒牛肉

做法

1. 牛腱肉、蕃茄和洋蔥洗淨切塊，老薑拍碎、蔥切段，蒜切碎。
2. 起油鍋放入洋蔥，以少許油炒軟，加入牛腱肉、蕃茄、蔥薑蒜、調味料和啤酒。
3. 煮滾後轉小火慢燉約2小時即成。

■ 一鍋配飯配麵都好吃的紅燒牛肉，散發著麥芽的香甜，其實做起來相當簡單吧！

材料：

牛腱肉	2條（約2斤）
黑啤酒	1瓶（600ml.）
蕃茄	2個
洋蔥	1粒
老薑	1塊
蒜碎	2大匙
蔥	5支

調味料：

冰糖	1大匙
醬油	1/2碗

Methods:

1. Rinse beef shanks, tomatoes and onion separately, then cut them into pieces. Crush old ginger and cut scallion strings into sections.
2. Heat a little oil in wok and stir-fry onion until soft. Add beef tendons, tomatoes, scallion sections, crushed ginger and minced garlic along with all the seasonings and beer.
3. Bring to a boil, then reduce heat to low. Continue simmering on low for about 2 hours until done. Ready to serve.

■ Serve with rice or noodles. This delicious simmered beef is brimming with the sweetness of maltose. In fact this dish is easy to prepare!

Ingredients:

2 beef shanks (approximately 2.6 lbs.), 1 bottle black beer (600ml.), 2 tomatoes, 1 onion, 1 piece old ginger, 2T minced garlic, 5 scallion strings

Seasonings:

1T rock sugar, 1/2 bowl soy sauce

Beef Steak with Beer
啤酒醃牛排

做法

1. 牛排洗淨，洋蔥洗淨切碎。
2. 醃料攪拌均勻，放入牛排浸泡約2小時。
3. 烤箱預熱至170℃，放入烤箱烤5分鐘後翻面續烤5分鐘即可享用。
4. 也可起油鍋，以小火將牛排煎成自己喜歡的熟度。

■ 如果你不習慣吃毫無調味的純美式牛排，不妨試試這道可口的啤酒醃牛排。

材料：

牛排　2塊（約1吋厚）

調味料：

新鮮檸檬汁　2大匙
鹽　適量
蒜碎　1大匙
洋蔥　1個
巴西里、奧勒岡、茵陳蒿末　各1/4小匙
啤酒　1杯（240 ml.）

Methods:

1. Rinse beer well. Rinse onion well and chop finely.
2. Combine marinade well together and soak steaks in marinade for about 2 hours until the flavor is absorbed.
3. Preheat oven to 170C and roast steaks for about 5 minutes first, then turn them over and continue roasting for 5 minutes longer. Remove and serve.
4. Or heat oil in pan and fry the steaks over low heat until they reach the texture you prefer.

■ If you are not used to the American-style steak, try this delicious steak with beer.

Ingredients:

2 steaks (approximately 1 inch thick)

Marinade:

2T. fresh lemon, salt as needed, 1T. minced garlic, 1 onion, 1/4 t. of parsley, oregano and tarragon each, 1C beer (240ml.)

Garlic Roasted Lamb Ribs
蒜烤羊排

做法

1. 羊小排洗淨，以刀背輕拍肉面，將纖維拍散較易入味。
2. 醃料攪拌均勻，放入羊排浸泡約1小時。
3. 烤箱預熱至170℃，放入烤箱烤5分鐘後翻面續烤10分鐘即可享用。

■ 超市即可買到羊小排，肉質鮮嫩，是燒烤的上好食材。蒜香四溢、口感滑嫩，保證不敢吃羊肉的人也忍不住想吃一口喔！

材料：

羊小排	4片

醃料：：

蒜碎	2大匙
鹽	1小匙
醬油	1小匙
果糖	1大匙
啤酒	100 ml.
黑胡椒	1/2小匙

Methods:

1. Rinse lamb ribs well, tenderize the meat fiber with the flat of the knife so that the flavor will be absorbed more easily.
2. Combine the marinade seasonings well together and soak lamb ribs for about 1 hour.
3 Preheat oven to 170C and roast lamb ribs for 5 minutes. Then turn over and continue roasting for 10 minutes until done. Serve.

■ Lamb ribs can be found at any supermarket. The texture of the meat is fresh and tender, it is a good choice for roasting. This dish is filled with garlic flavor and the texture is smooth as well as tender, even people who don't eat lamb would certainly want to have a bite!

Ingredients:

4 slats lamb short ribs,

Marinade:

2T. minced garlic, 1t. salt, 1t.soy sauce, 1T. fructose, 100ml. beer.1/2t black pepper

Bar

Mustard Ham with Beer
啤酒芥末火腿

做法

1. 烤箱預熱至170℃，放入火腿烤10分鐘後取出。
2. 醃料攪拌均勻，先刷一半的份量於火腿表面，放入烤箱續烤10分鐘。
3. 取出火腿，再將剩餘的醃料均勻塗上，入烤箱烤10分鐘即可取出切片食用。

■ 鳳梨切片，或如圖所示搭配火腿吃。有著濃濃的夏威夷風味！

材料：

市售熟火腿1個(約1,000公克)
鳳梨 1/4個

醃料：

黑糖 3/4杯
法式芥末醬 2大匙
黑啤酒 1/2杯(120 c.c.)

Methods:

1. Preheat oven to 170C and roast ham for 10 minutes, then remove.
2. Combine the marinade well together. Brush half of the portion over the surface of the ham, then return to oven and continue roasting for 10 minutes longer.
3. Remove ham and brush evenly with the remaining marinade. Return to oven once again for 10 more minutes until done. Remove and cut into slices. Serve.

■ Cut pineapple into slices. Or serve with ham as the photo shows. This dish is filled with Hawaiian flavor!

Ingredients:

1 market-sold cooked ham (approximately 2.2 lbs.), 1/4 pineapple

Marinade:

3/4C dark brown sugar, 2T. French mustard, 1/2C black beer (120c.c.)

Apricot-Filled Pork Rib Chop

烤甜杏夾心豬排

做法

1. 排骨肉中間橫切一刀呈開口袋狀。
2. 將杏乾浸泡在啤酒中3小時，取出後夾入排骨肉的切口中，以棉線將肉排捆緊成圓筒狀，表面抹上鹽和胡椒。
3. 取一深底鍋將肉排兩面煎黃，倒入浸過杏仁乾的啤酒及高湯，以小火燜煮3小時即成，注意快乾時要加高湯。
4. 食用時取下棉線，將肉排切厚片排盤，淋入鍋中剩餘的肉湯。

■ 肉排吸收了杏乾的果香及麥芽的甜香，是道老少咸宜的下飯菜。

材料：

去骨排骨肉	1條（約1斤）
杏乾	20個
啤酒	1杯
高湯約	3杯

調味料：

鹽、白胡椒	適量

Methods:

1. Make a slit horizontally in the center of the rib to form a pocket.
2. Soak apricots in beer for 3 hours until flavor is absorbed. Remove apricots and retain the beer for later use, then stuff inside the rib pocket. Tie tightly with cotton string into a cylinder and coat the surface evenly with salt and pepper.
3. Fry the pork cylinder in a deep frying-pan until golden on both sides. Add the beer from method (2) along with the soup stock. Cover and simmer over low heat for 3 hours until done. Add a little soup stock if the liquid starts drying out.
4. Remove and discard the cotton string, cut the stuffed pork ribs into thick slices. Arrange in order in serving plate and drizzle with the liquid from the pan. Serve.

■ Pork ribs that have absorbed the fruity aroma of the apricot and sweet flavor of the maltose are a wonderful appetizer for either young and old.

Ingredients:

1 slat boneless ribs (approximately 1.3lbs.), 20 apricots, 1C beer, approximately 3C soup stock

Marinade:

salt and white pepper as needed

White Sauce Meatballs with Beer
啤酒白汁肉丸

做法

1. 九層塔洗淨切碎，加入絞肉中，與鹽、胡椒、太白粉拌勻成肉丸。
2. 起鍋，放入奶油和麵粉炒香，加入啤酒和高湯炒成濃稠狀。
2. 放入肉丸，以中火將肉丸煮熟。起鍋前可再加適量的鹽和白胡椒粉調味。

■ 肉丸好吃，醬汁也非常美味，可澆在義大利麵或白飯上享用，搭配麵包也很適合！

材料：

絞肉	1斤
九層塔	2大匙
啤酒	1/2杯
高湯	1杯

調味料：

鹽	1大匙
白胡椒	1/2小匙
太白粉	1大匙
奶油	2大匙
麵粉	4大匙

Methods:

1. Rinse basil well and chop finely. Add to ground pork along with salt, pepper and cornstarch added. Mix well and make into meatballs.
2. Heat wok, stir-fry butter and flour until fragrant. Pour in beer and soup stock, then stir until the soup has thickened.
3. Add meatballs and cook on medium until the meatballs are done. Season with salt and white pepper as needed to taste before remove from heat. Serve.

■ Meatballs are delicious, and so is the sauce. It can also be served by drizzling it over pasta or white rice, and goes well with bread.

Ingredients:

1.3 lbs. ground pork, 2T. basil, 1/2C beer, 1C soup stock

Seasonings:

1T. salt, 1/2t. white pepper, 1T. cornstarch, 2T. butter, 4T. flour

Red-Cooked Pork Feet
紅燒豬腳

做法

1. 豬腳洗淨切塊、蔥切段。
2. 起油鍋加入1小匙油和砂糖，以小火加熱，不停攪拌直到砂糖融解成棕色，浮現很多泡沫。
3. 加入豬腳塊不停翻炒，讓豬腳塊均勻沾滿棕色的糖汁；倒入醬油小滾片刻，此時豬腳呈現漂亮的棕黃色澤。
4. 倒入蔥薑蒜、啤酒、花椒八角和辣椒，以小火燜煮約2小時，至豬腳已爛但仍有韌的口感。

■ 嘗一嘗不同風味的豬腳，清淡而鮮美。

材料：

豬腳	1隻（約2斤）
啤酒	1瓶（600ml.）
老薑	5片
蒜頭	10瓣
蔥	3支

材料2：（過程中添加）

砂糖	3大匙
醬油	1/2碗
花椒八角	少許
辣椒	隨意
油	1小匙

Methods:

1. Rinse pig's foot well and cut into pieces. Cut scallions into small sections.
2. Heat 1 small spoonful of oil and granulated sugar over low heat in pan. Stir constantly until the sugar dissolves and becomes brown with lots of bubbles floating on surface.
3. Add pig's foot pieces and stir thoroughly and constantly to ensure the pig's foot pieces are coated with brown sugar syrup. Pour in soy sauce and bring to a boil for a minute to two. The pig's foot pieces should appear a beautiful golden brown color.
4. Add scallion, ginger slices and garlic cloves as well as beer, peppercorns, star anises and chili pepper to mix. Simmer over low heat for about 2 hours until the pig's foot pieces are tender and springy.

■ Try this different flavored, light, and delicious beer cooked pig's foot.

Ingredients:

1 pig's foot (approximately 2.6 lbs.), 1 bottle beer (600ml.), 5 ginger slices, 10 cloves garlic, 3 scallions

Seasonings:

3T. granulated sugar, 1/2 bowl soy sauce, Schezwan peppercorns and star anises as needed, chili peppers as desired

Fried Cod Steak
煎鱈魚排

做法

1. 鱈魚排洗淨，放在深盤中。
2. 醃料攪拌均勻，放入鱈魚排浸泡約30分鐘。
3. 起油鍋，以1大匙橄欖油或葡萄籽油小火煎成金黃色即成。

■ 鱈魚含有豐富的魚油，所以不必放太多油，這道菜也可以換成鮭魚。

材料：

鱈魚排	約1斤

醃料：

新鮮檸檬汁	1大匙
鹽	1大匙
啤酒	100 ml.

Methods:

1. Rinse cod steak well and remove to a deep plate.
2. Combine the marinade well and soak cod steak in marinade for about 30 minutes.
3. Heat pan, add 1T. of olive oil or grape seed oil until smoking, fry cod steak over low heat until golden and remove. Serve.

■ Cod contains abundant fish oil. It is not necessary to add lots of oil. Substitute salmon for the cod if desired.

Ingredients:

cod steak approximately 1.3 lbs.

Marinade:

1T. fresh lemon juice, 1T. salt, 100ml. beer

Steamed Grass Shrimp
清蒸草蝦

做法

1. 草蝦洗淨去腸泥。
2. 將草蝦與鹽、花椒、啤酒一起放入大盤中，以大火蒸約7分鐘即可趁熱享用。

■ 鮮嫩的蝦肉、淡淡的花椒香和麥芽香，實在是人間美味。

材料：

草蝦	1斤

材料2：（過程中添加）

鹽	1大匙
花椒	數粒
啤酒	200 ml.

Methods:

1. Rinse shrimp well and devein.
2. Place shrimp on a large plate along with salt, peppercorns and beer added. Steam in steamer on high for about 7 minutes until done. Remove and serve while still steaming.

■ With fresh tender shrimp meat, light peppercorn and maltose aroma, this is really a heavenly dish.

Ingredients:

1.3 lbs. grass shrimp

Seasonings:

1T. salt, Szechwan peppercorns as needed, 200ml. beer

Beer Shrimp
啤酒蝦

做法

1. 蝦洗淨去腸泥，放入蒸盤備用。
2. 起油鍋，以2大匙橄欖油爆香大蒜、巴西里，加入鹽調味，倒入啤酒煮滾後立即熄火，倒入**1**中。
3. 以大火蒸熟即可趁熱享用。

■ 新鮮的蝦肉要剛熟時馬上入口最好吃，否則蝦肉縮了就不好吃了，這道啤酒蝦，讓你吃到鮮蝦的甘美。

材料：

新鮮草蝦或海蝦	1斤
大蒜碎	1大匙
巴西里	3大匙
啤酒	60 ml.

調味料：

鹽	適量

Methods:

1. Rinse shrimp and devein,put into a deep bowl.
2. Heat pan and add 2T. of olive oil until smoking. Stir-fry minced garlic and parsley until fragrant. Season with salt to taste. Pour in beer and bring to a boil, then remove from heat,pour onto **1**.
3. Steam over high heat until done. Serve right away when the dish is still steaming.

■ Fresh-cooked shrimp tastes the best. If not fresh-cooked, the meat will shrink and lose its flavor. You can taste the freshness in this beer steamed shrimp dish.

Ingredients:

1.3 lbs. fresh grass shrimp or sea shrimp, 1T. minced garlic, 3T. parsley, 60ml. beer

Seasonings:

salt as needed

Beer Pine Nut Dressing
啤酒松子醬

做法

1. 松子烤熟。
2. 全部材料放入食物處理機或果汁機打碎即成。

■ 澆在生菜沙拉上或是塗法國麵包、拌義大利麵都很好吃。

材料：

九層塔碎	2大匙
松子	2大匙
香菜碎	1大匙
蒜碎	1大匙
橄欖油	60ml.
啤酒	60ml.
紅酒醋	2大匙
鹽	1小匙

Methods:

1. Roast pine nuts until done and remove.
2. Combine all the ingredients in food processor or blender, and beat until fine.

■ Drizzle over salads or spread over the French bread. It is very good with spaghetti.

Ingredients:

2T. chopped basil, 2 T. pine nuts, 1T. chopped cilantro, 1T. minced garlic, 60ml. olive oil, 60ml. beer, 2T. red wine vinegar, 1t. salt

Cold Mixed Napa Cabbage with Vinegar and Beer

涼拌白菜心

做法

1. 大白菜洗淨去除外面老葉，取嫩白菜心切細絲。滷豆干、青蔥、辣椒切細絲。
2. 蒜茸花生稍壓碎。
3. 全部材料放在大碗內，充分拌勻即可食用。

■ 這道菜要現做現吃，否則白菜碰到鹽，時間久了會出水，就不好吃了。

材料：

大白菜	1棵
香菜末	2大匙
滷豆干	2塊
蒜茸花生米	3大匙
青蔥	少許
辣椒	少許

調味料：

鹽	1小匙
糖	1小匙
醬油	1/2小匙
麻油	1大匙
米醋	50 ml.
啤酒	50 ml.

Methods:

1. Discard the outer old leaves and retain the tender napa cabbage heart for shredding. Shred pressed tofu, scallions and chili peppers finely.
2. Crush peanuts slightly.
3. Combine all the ingredients with all seasonings in a large bowl. Mix well completely and serve.

■ This dish has to be served right away after it is prepared, or the cabbage will become soggy once it has been salted. Then the dish will not be as good.

Ingredients:

1 head napa cabbage, 2T. minced cilantro. 2 pieces stewed pressed tofu, 3T. garlic-flavored shelled peanuts, scallions and chili peppers as needed

Seasonings:

1t. salt, 1t. sugar, 1/2t. soy sauce, 1T. sesame oil, 50ml. rice vinegar, 50ml. beer

Stir-Fried Kale

炒青菜

做法

1. 青菜洗淨，蒜頭拍碎。
2. 起油鍋，加入1大匙油爆香蒜頭，倒入芥蘭菜及玉米筍快炒，加入鹽和啤酒稍拌炒即可起鍋。

■ 綠油油的青菜以大火快炒，色澤翠綠、配上淡淡酒香，好看又好吃。

材料：

芥蘭菜	1斤
玉米筍	6支
蒜頭	2瓣

調味料：

鹽	少許
啤酒	50 ml.

Methods:

1. Rinse kale well. Crush garlic cloves.
2. Heat wok and add 1T. of cooking oil until smoking. Stir-fry garlic cloves until fragrant. Add kale and baby corn to mix. Season with salt and beer to taste. Mix well rapidly and remove from heat. Serve.

■ Green fresh vegetables sauteed over high heat. With its beautiful green color and light beer flavor, this dish is not only good looking, but also delicious.

Ingredients:

1.3 lbs. Chinese kale, 6 ears baby corn, 2 cloves garlic

Marinade:

salt as needed, 50ml. beer

San Francisco Seafood Soup

舊金山海鮮湯

做法

1. 螃蟹洗淨切塊、蝦洗淨去腸泥、其他海鮮均洗淨。洋蔥洗淨切大塊。蕃茄洗淨切碎。
2. 起油鍋，以3大匙橄欖油炒香洋蔥，加入除海鮮外的全部材料，以小火燜煮1小時。
3. 放入海鮮煮10分鐘，加適量鹽調味即可趁熱享用。

■ 在舊金山漁人碼頭，點一客海鮮湯配上獨特的酸麵包(Sourdough Bread)，佐著空氣中微濕的海洋味，才讓旅人們有不虛此行的感覺。

■ 從舊金山友人處學會了這道菜的配方，我又添加上了啤酒。每年的元旦大餐，我就煮這麼一鍋，讓全家大小吃個飽。

材料：

新鮮螃蟹	2斤
草蝦或海蝦	1斤
新鮮干貝	500克
大蛤蜊	10個

調味料：

蒜頭	2瓣
洋蔥	1個
蕃茄	2個
蕃茄糊	1小罐（6oz.）
月桂葉	2片
百里香碎、奧勒岡碎各	1/2小匙
黑胡椒	1/2小匙
啤酒	1瓶
白蘭地	1/4杯（60ml.）
高湯	2杯（480 c.c.）

Methods:

1. Rinse crabs well and cut into pieces. Rinse shrimp and devein. Rinse the remaining seafood well. Rinse onion and cut into large chunks. Rinse tomatoes and chop finely.
2. Heat 3T of olive oil in pan and stir-fry onions until fragrant. Add all the ingredients except seafood and simmer over low heat for 1 hour.
3. Add seafood and continue cooking for 10 minutes. Season with salt as needed. Serve while still steaming.

■ On Fisherman's Wharf in San Francisco, order a dish of seafood soup with its unique sourdough bread. With the humid, salty ocean breege in the air, it makes travelers feel the trip was worth it.

■ I learned this recipe from a friend from San Francisco and I added beer to it. Every year for the New Year meal, I prepare a large pot of it and fill every belly in the family with it.

Ingredients:

2.6 lbs. fresh crabs, 1.3 lbs. grass shrimp or sea shrimp, 500 grams fresh scallops, 2 cloves garlic, 10 large clams, 1 onion, 2 tomatoes, 1 small can tomato paste (6 oz.)

Seasonings:

2 bay leaves, chopped thyme, 1/2t. chopped oregano, 1/2 t. black pepper, 1 bottle beer, 1/4 C (60ml.) brandy, 2C soup stock (480c.c.)

懶人啤酒

照著配方輕鬆做，在家輕鬆喝啤酒

材料：

濃縮麥芽汁1罐
乾燥麥芽粉1包
啤酒花顆粒1包
酵母菌1包
吉利丁1包
果糖3/4杯（180ml.）
成品份量：23瓶 × 0.6公升
製作時間：4週

前製作業

取一洗淨的塑膠桶，放入4公升的水及60公克的偏亞硫酸鉀，充份溶解後把要用的工具，如長柄匙、發酵鎖、虹吸管、濾網、比重測糖計等，全部泡在裡面。

做法

1. 不銹鋼大鍋s中倒入6公升的水煮開，加入濃縮麥芽汁繼續煮沸15分鐘，熄火前放入啤酒花顆粒、加入麥芽粉，攪拌均勻。
2. 將10公升冰開水倒入玻璃發酵瓶中，將麥芽汁過濾倒入。
3. 取出半杯麥芽汁放涼（約16°C），以比重測糖計量取糖的含量，紀錄在筆記本上。
4. 將酵母菌加入半杯溫水（約38°C）中活化。
5. 當麥芽汁降溫到24°C時，加入已活化的酵母菌，攪拌均勻。放置在24°C左右的陰暗處發酵。
6. 24小時後檢查發酵作用是否開始，發酵鎖有氣泡不斷的湧出表示發酵正常。
7. 約一週後，當發酵鎖湧出的氣泡逐漸減弱而停止時，再倒出一些麥芽汁以比重測糖計檢測含糖量，降到0度左右時即表示第一次發酵作用完成，虹吸換瓶。

第二次發酵及澄清處理

8. 將兩杯水與3/4杯果糖一起放入鍋中煮開。
9. 吉利丁溶解於300 ml.冷水中，隔水加熱成透明狀。
10. 糖水和吉利丁液倒入新釀啤酒中，以長柄匙攪拌均勻。
11. 將啤酒瓶浸泡在偏亞硫酸鉀水溶液消毒後，再用熱水裡外沖洗掉偏亞硫酸鉀水。
12. 利用虹吸管，將新釀啤酒裝滿啤酒瓶，不要裝的太滿，餘留大約2吋的空間。加蓋密封。
13. 放置在6°C左右陰暗的地方，熟成2週。

超簡單的啤酒組合包

Beer Brewing Package

為了照顧讀者的需求，本社特別自國外進口釀製啤酒必備的原料，組合成一套「超簡單啤酒組合包Beer Brewing Package」，有了組合包，你就不必傷腦筋張羅材料，照著食譜，輕鬆製作出懶人啤酒。組合包內容物如下：

1

2 3

1.濃縮麥芽汁1罐（可選擇黑啤酒或琥珀啤酒）
2.乾燥麥芽粉1包
3.啤酒花顆粒1包
4.酵母菌1包
5.吉利丁1包
6.偏亞硫酸鉀1包
7.發酵鎖1個
8.比重測糖計1個
9.20公升玻璃果酒桶（已鑽孔，可放入發酵鎖）
10.虹吸管一根
11.啤酒蓋

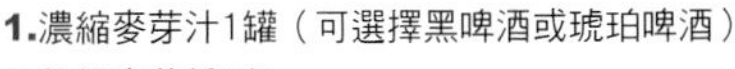

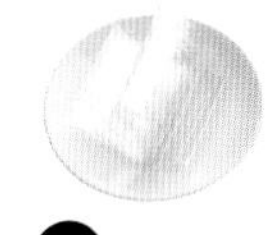

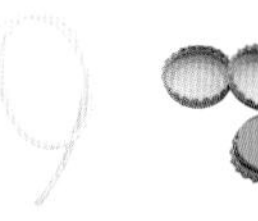

4 5 6 7 8 9 10 11

「超簡單啤酒組合包Beer Brewing Package」
可釀出23瓶0.6公升啤酒，
特惠本書讀者，有興趣者請電(02)2345-1958

國家圖書館出版品預行編目資料

在家釀啤酒Brewers' Handbook——啤酒DIY和啤酒做菜／錢 薇著. -- 初版. -- 台北市：朱雀文化, 2005[民94]

面；公分. -- (Cook50；54)

ISBN 986-7544-28-5（平裝）

1.酒 - 製造　2.食譜

463.82　93023217

在家釀啤酒

啤酒DIY・啤酒做菜

COOK50054

作　者■錢 薇　攝　影■徐博宇、林宗億　食譜編輯■任 興　美術編輯■許淑君

企畫統籌■李 橘　發行人■莫少閒　出版者■朱雀文化事業有限公司

地　址■台北市基隆路二段13-1號3樓　電話■(02)2345-3868　傳真■(02)2345-3828

劃撥帳號■19234566 朱雀文化事業有限公司　e-mail■redbook@ms26.hinet.net

網　址■http://redbook.com.tw　總經銷■展智文化事業股份有限公司

ISBN■ 986-7544-28-5　初版一刷■2005.01

定　價■320元　出版登記■北市業字第1403號